EXPERIMENTS IN PHYSICS
PRINCIPLES AND APPLICATIONS
FOURTH EDITION

Norman C. Harris
Professor Emeritus of Higher Education
The University of Michigan

A. James Mallmann
Professor of Physics
Milwaukee School of Engineering

GLENCOE
McGraw-Hill

New York, New York   Columbus, Ohio   Woodland Hills, California   Peoria, Illinois

CONTENTS

Experiments in Physics, Fourth Edition

Imprint 2002 Copyright © 1990 by Glencoe/McGraw-Hill. All rights reserved. Copyright © 1990 by McGraw-Hill, Inc. All rights reserved. Copyright © 1980, 1972, 1963 as *Experiments in Applied Physics* by McGraw-Hill, Inc. Printed in the United States of America. Except as permitted under the United States Copyright Act, no part of this publication may be reproduced or distributed in any form or by any means, or stored in a database or retrieval system, without prior written permission from the publisher. Send all inquiries to: Glencoe/McGraw-Hill, 8787 Orion Place, Columbus, Ohio 43240.

5 6 7 8 9 10 11 12 13 14 15   045   07 06 05 04 03 02

ISBN 0-07-026852-5

PREFACE

At the beginning of the twentieth century there was general agreement, even among scientists, that most of the important discoveries in physics had been made. New applications of basic general principles would follow, of course, and minor refinements in established theories might occur, but the consensus was that everything of real significance had already been discovered and quite satisfactorily explained. In summary, pre-twentieth-century physics, now known as "classical physics," had posed all the important problems and arrived at all the answers.

How premature that judgment was! Since 1900, experimental and theoretical research in the fields comprising "modern physics" has created an entirely new world of science and technology, including electronics and its spinoffs into television, telecommunications, space exploration, computers, and medical technology; fiber optics and lasers; radar, sonar, and microwave technology; solar energy and photovoltaics; high-energy (particle) physics; magnetic resonance imaging; and nuclear energy. These fields of modern physics have spawned billion-dollar industries and created the so-called "high tech" society we now live in, and the pace of new discoveries and innovation is undiminished. One recent example is typical—new discoveries in electrical superconductivity since 1986 suggest the possibility that a century's knowledge of and development in electricity and magnetism may soon be due for major revision.

Theories are often formulated from mathematical analysis and then checked by careful laboratory tests. Just as often, a theory may result from the findings of laboratory research carried out for months or even years. Theoretical research is quickly followed by applied research in laboratories operated by universities, corporations, or government agencies. It is applied research that turns theoretical physics into applied physics, and laboratory work is the indispensable component of that process.

This fourth edition of *Experiments in Physics* will give students some idea of the scope of physics, and will provide opportunity for actual experience in some of the elementary techniques of applied research. Students will learn some of the basic principles of research methods and will have an opportunity for actual experience with laboratory equipment of considerable precision.

Physics is, by necessity and tradition, a laboratory science. Students cannot learn to apply physical principles without engaging in the "hands-on" use of scientific equipment and instruments. Demonstrated lectures, with the instructor performing all the steps, and with the equipment preset to yield the "correct" answer, can be instructive up to a point, but they provide little insight into the painstaking nature of laboratory research and they give the student no indication of his or her aptitude for or interest in a career in science, engineering, or technology.

A proper guide for laboratory work in physics will point the way for the student, assisting him or her in understanding the principles and in attacking the problem to be studied experimentally. However, the responsibility for *thinking through the problem* should be placed on the student. This manual attempts to be that kind of guide, and it places emphasis on the necessity for student thought and student analysis. An *original* written report of each laboratory investigation is assumed. Suggestions are provided to aid in setting up equipment, in data collection, and in data analysis, but the laboratory report itself is to be an original product of the student's own thought and ingenuity.

One important objective of laboratory work in physics is to train students in the art and mechanics of technical report writing. The laboratory reports should be regarded as though they were reports of a technical investigation in an actual job situation. A suggested format for laboratory reports is provided in the "Introduction to Students" section, but instructors may want to modify it to suit their own requirements for report writing.

In the present edition greater emphasis has been placed on the SI-metric system of units, although English system units have been retained for many of the experiments. The mks system has not yet replaced the English system in United States industry and commerce, and it is absolutely necessary that engineers and technicians be thoroughly competent in *both* systems. The experiments in this manual will provide ample opportunity to develop that competence. Any basic applied physics text covers systems of measurement and their units. *Physics: Principles and Applications,* for example, provides an excellent treatment of both systems, with a full explanation of the units used in basic physics and in engineering and technology.

Many of the experiments are similar to those found in general physics laboratory manuals, and the apparatus specified or recommended is typical of that found in most college physics laboratories equipped for first-year courses. Photographs and sketches included with each experiment will assist instructors in planning the laboratory exercises and in providing suitable apparatus.

Fifty-five experiments are included, from which instructors may choose those to be assigned as "required." If an open-lab concept prevails, some students may elect to do several more than the thirty-odd required exercises. If time constraints permit, it is strongly recommended that the apparatus and equipment *not* be neatly set up in advance for the students. Students should come to the laboratory session having studied the experiment in advance, and should be allowed the experience of setting up the apparatus, making necessary adjustments, experiencing trial and error (with due regard for expensive equipment and safety, of course), and perhaps getting wrong or inconclusive results occasionally, requiring re-analysis and repetition of trials and measurements.

Suggestions from many users of earlier editions of the manual have been received, and most of them have been incorporated in this revision. Instructors from technical institutes, community colleges, and state colleges in every region of the country have been most helpful, and their advice has resulted in modifications and improvements to many of the experiments. Several experiments from prior editions have been omitted as a result of these suggestions, and four new experiments have been added. New topics include calibration of a thermometer, radiant heat transfer, the P-N junction diode, and the integrated-circuit timer.

Many new sketches have been provided to assist students in the analysis and setup of experiments. Photographs have been updated throughout, suggesting the types of laboratory equipment available from commercial suppliers.

ACKNOWLEDGEMENTS
The authors are grateful to the several "high-tech" industrial and research firms that provided an opportunity to observe at first hand both the layout and the equipment of modern research laboratories, and the work of technicians in basic, engineering, and medical research. Sincere appreciation is expressed for the cooperation of laboratory apparatus suppliers in furnishing photos and specifications for the equipment suggested herein. Their contributions are generally identified in the legends for the photos and illustrations. Special thanks are due to Academic Media Services, Milwaukee School of Engineering, for photographic services; and to Mr. David Watson for his photographic skills.

NORMAN C. HARRIS
A. JAMES MALLMANN

INTRODUCTION TO STUDENTS

PURPOSE OF LABORATORY WORK

The purpose of laboratory work in physics is to provide an opportunity for you to learn how physics applies to modern life by your own participation, observation, testing, problem solving, and thinking. An understanding of applied physics cannot be achieved entirely from reading the textbook or from attendance at demonstrated lectures. Many basic principles and their applications can be understood only by actual participation in experimental investigations designed to illustrate the principle and to suggest applications. A second and equally important purpose is to acquaint you with laboratory equipment and procedures, and to give you an opportunity to work carefully with precision equipment. Work in the physics laboratory is an introduction to the work of research and test technicians in industry, engineering, allied health, agriculture, and business. Your interest in and aptitude for physics laboratory experiments may be a good indication of your suitability for a career in science, engineering, or technology.

PRECISION EQUIPMENT

Experience in handling scientific instruments and precision equipment is of unquestioned value to technicians in all fields, since the controlled processes of modern industry, business, and medicine require instruments and equipment of extreme sensitivity and precision.

LABORATORY REPORTS

A very important phase of laboratory work is the laboratory report itself. Success in the professions or in managing one's own business depends on the ability to explain to others the events, processes, plans, and problems which are dealt with day by day. Often these sessions involve test data, equipment analysis, research reports, computer calulations, and written recommendations. Writing laboratory reports in physics is excellent training for such responsibilities. Preparation of lab reports for your weekly lab experiments will sharpen your powers of analysis and improve the clarity of your written communication.

STUDY IN ADVANCE

Laboratory experiments are usually performed weekly. The laboratory exercise assigned by the instructor should be thoroughly studied *before you come to the laboratory session*. Reading in the text and in the references assigned should be faithfully performed so that there will be no delay in getting started on the experiment in the laboratory period. Advance preparation is especially important if your lab instruction is conducted in an "open lab" setting, where you might be working alone or with a group, without direct instructor supervision.

PERFORMING THE EXPERIMENT

Check the apparatus provided, and ascertain at once if there are any shortages or malfunctions. Set up the equipment in accordance with instructions. Proceed carefully and develop scientific, methodical work habits. Remember that most scientific equipment is *very expensive* and quite susceptible to damage. If the setup is at all complicated, ask the instructor or the lab assistant to inspect your layout before you proceed with the actual performance of the experiment. If electricity, compressed air, hot water, steam, or operating machinery are involved, be sure to *observe all safety rules*.

Take the measurements required, recording them neatly in tabular form on the Data Sheet, which is intended to become a part of your laboratory report. Note that it is perforated for easy removal. Make a sketch of the equipment layout for your reference later, as you write up the report. Double-check to make sure that you have recorded all necessary data.[1]

Upon completion of the experiment, carefully disassemble the apparatus and replace the equipment as directed by your instructor.

WRITING THE LABORATORY REPORT

Laboratory reports are due on a definite schedule set up by the instructor. Late reports are ordinarily subject to penalty, so form at once the work habit of completing reports on time! They should be either typewritten or handwritten in ink on one side of the paper only. Careful attention should be given to penmanship, general layout, spelling, and grammar. Your instructor may have a definite format which is required for your laboratory report. If so, follow it to the letter. If not, the suggested format shown on the next page may be useful to you, at least in the beginning weeks of the work.

LABORATORY NOTEBOOK

Laboratory reports will ordinarily be evaluated and returned to you on schedule. You should build a *laboratory notebook* (under separate cover) from them as they are returned. Strive to make each report an improvement over the preceding one. Your laboratory notebook will be evidence of your gradually increasing ability to investigate and discuss intelligently the scientific principles which are essential to engineering, technology, industry, business, and medicine.

---

[1]At first the Data Sheets provided in this manual will be quite complete, for your guidance; but in later experiments they may be suggestive only, or omitted entirely, so that you can develop skill in providing your own data-recording system.

SYSTEMS OF MEASUREMENT AND UNITS

Both the SI-metric system and the English system of measurement will be used in this manual. The reason for working with both systems is that, in the United States, both systems are in use by engineers, technicians, and researchers as they work in industrial, business, aerospace, allied health, and agricultural jobs.

Please refer to your textbook for a complete discussion of systems of measurement and for definitions of units. Some texts deal only with the SI-metric (meter-kilogram-second) system and barely mention the English (foot-pound-second) system common in the United States. The text *Physics: Principles and Applications* deals fully with both systems of measurement and explains how to use them in a variety of situations and problems. (See Chapters 2 and 5 of that text.)

The pound is used in this manual only as a unit of *force* or *weight*. The kilogram is the SI-metric unit of *mass:* 1 kilogram (kg) = 1000 grams (g). The newton (N) is the basic unit of force and weight in the metric system. For further clarification of force and mass units, see Experiment 2.

Some of the experiments in this manual will use the English system, while others will use the SI-metric system. If your instructor wishes you to use a different system of measurement or different units from those for which the experiment is written, change the Data Sheet accordingly.

PLOTTING AND INTERPRETING CURVES AND GRAPHS

In engineering, science, business, and medicine there are many instances where it is advantageous to portray graphically the manner in which two variables are related. As a simple example, consider the relationship between the volume of a gas and its absolute pressure when the temperature remains constant. Boyle's law of gases (to be studied in one of the experiments to follow) states that for a given mass of gas held at a constant temperature, the product of the pressure times the volume is constant. The equation $PV = k$ expresses this relationship. (The letter k is often used in physics to denote a constant.)

Measurements of P and V may be plotted on graph paper, and a curve obtained. The nature of the curve has mathematical significance and, in addition, assists the student to grasp the actual physical significance of this important law of gases. Many other important physical laws, generalizations, and concepts can best be understood from a graphical plot like the P-V diagram in Fig. A.

Graph paper should be used for all curve plotting. The graph paper used should allow for decimalization in plotting and reading; that is, the spaces between heavy rules on the paper should be divided into ten (or five) spaces by fainter lines. Draw horizontal and vertical axes along lines well inside the margin of the paper. The quantity which is

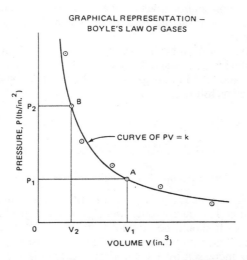

Fig. A. Graphical representation of Boyle's Law of Gases. Pressure-volume relations in a given masss of gas that is held at a constant temperature. The area of the rectangle P_1AV_10 equals area of rectangle P_2BV_20; or, $P_1V_1 = P_2V_2$. For all values of pressure, volume varies in such a manner that the product of pressure and volume is constant, or $PV = k$.

considered the *independent variable* is usually scaled off (plotted) along the horizontal axis. These values are called *abscissas*. Corresponding values of the *dependent variable* are laid off to scale along the vertical axis. These are called *ordinates*. The quantity whose variations ("ups and downs") are of the most interest is usually plotted along the vertical axis.

Choose a scale for both axes such that the plot will occupy nearly all of the remaining portion of the coordinate paper. It is not necessary that the same scale be used for both axes. Take as the origin the intersection of the axes near the lower left-hand corner of the sheet.

Label each axis with the quantity (and units) it represents, and indicate the scale used. Each pair of values is then plotted carefully and designated by a point with a small circle drawn around it. Some "curves" are of the broken-line type, obtained by connecting the plotted points with straight lines. Other physical relationships are best represented by "smoothed" curves, fitted to the general contour of the plotted points. Such curves do not actually pass through all the plotted points, but are an attempt to portray results from the study of a law or principle in such a way as to disregard random errors (see Fig. A).

Every graph must be given a title. A brief interpretation of the significance of the graph should also be included. All identifying information on a graph should be neatly *printed*, and neatness throughout should be emphasized. If more than one curve is plotted on a single set of axes, the curves may well be drawn in different colors for clarity of interpretation. Or, different plotting symbols (such as ⊙, △, and ⊡) can be used to plot each curve.

ERRORS IN LABORATORY WORK

No measurement is absolutely accurate. Errors creep into the most careful of investigations, even when equipment and instruments of high precision are used.

Errors in laboratory work are classified under two headings—systematic errors and random errors. *Systematic*

errors are the result of some fault in the system or equipment used. For example, using a poorly calibrated thermometer, such as one which reads 102 degrees C when the correct value is 100 degrees C, will introduce systematic errors into the experiment. Another example of a systematic error would be the results obtained using a micrometer caliper whose "zero" reading (spindle fully closed on the anvil) was 0.006 in. instead of 0.000 in.

Random errors can result from a variety of causes—for example, poor judgment or actual blunders on the part of the experimenter, improper use of an instrument or apparatus, carelessness in reading a dial or scale, or fluctuations in the laboratory electrical voltage, water pressure, etc.

PRECISION VERSUS ACCURACY

Scientists engaged in original research are interested in the precision of their work. By *precision* is meant the degree to which a series of measurements repeated under the same conditions will give results which agree closely with each other. From such original investigations come physical laws and the "accepted values" of physical constants which are found in handbooks and in tables such as those in the Appendix of this laboratory manual.

The precision of a set of measurements can be evaluated by first obtaining the simple arithmetic average (or mean) of all the trial measurements and then determining the *standard deviation* of the mean. The smaller the standard deviation, the more precise is the set of measurements. (See any elementary text on statistics for a method of calculating standard deviation.) Standard deviation is an indicator of the dispersion or "scatter" of a set of measured values.

In the elementary laboratory work of freshman college physics, however, we are not attempting to establish the value of a physical constant or to discover a physical law, but are merely trying to work with sufficient care to obtain a reasonably accurate check on a quantity, law, or principle whose correctness has long been established. The *accuracy* of a measurement is determined by how closely it checks with a previously determined standard, or "accepted value." For the purposes of this laboratory course, a satisfactory evaluation of the accuracy of experimental results may be obtained by first finding the arithmetic mean of the several trial results and then calculating the percent error of the mean from the accepted value. Percent error is calculated as follows:

Percent error =

$$\frac{\text{Experimental mean value} - \text{"correct" value}}{\text{"Correct" value}} \times 100$$

Accepted (correct) values may be found in handbooks, in standard reference works, and in your textbook, as well as in the Appendix of this book. In your lab report, cite the source of "correct values" that you use.

After calculating the percent error of your results, analyze the errors, pointing out to what extent they are systematic or random.

STANDARDS FOR RECORDING DATA AND MAKING CALCULATIONS

In recording measurements, all figures that can be trusted should be entered on the Data Sheet. For example, if a dial indicator is known to be accurate to the nearest ten-thousandth of an inch, readings with that instrument should be recorded, for example, as 0.0058 in., not rounded to 0.006 in. Conversely, a measurement should not be written down to imply a greater accuracy than is actually inherent in the measuring device. A reading of 1.056+ in. on a micrometer caliper designed to read to the nearest thousandth of an inch should not be "guessed at" and recorded as 1.0563 in. The reading 1.056 means that all the figures recorded can be trusted. If the reading were recorded as 1.0563, the implication would be that the 3 in the ten-thousandths place was also to be trusted, but this is contrary to fact.

A recorded figure of 1.05 means that the quantity can be relied on as accurate to three significant digits; a reading of 1.056 is said to be accurate to four significant digits. The position of the decimal point has no effect on the number of significant digits in a quantity. The numbers of 1.056 and 105.6 both have the same number of significant digits. Zeros which merely locate a decimal point are "place holders," not significant digits. But if a zero represents a value actually obtained from an instrument or a calculating device, then it is a significant digit. For example, in the number 0.00605, the zero to the left of the decimal point and the two zeros immediately to the right of it are not significant digits, but the zero between the 6 and the 5 is a significant digit.

Zeros must be used even with whole numbers if a stipulation of a certain degree of accuracy is to be made. For example, a measurement recorded as 68 g means that the balance used was one of low precision and that the value actually lies somewhere between 67.5 g and 68.5 g. If the mass of the object had been determined as exactly 68 g, accurate to the nearest hundredth of a gram, the value should have been recorded as 68.00 gm, and the two zeros to the right of the decimal point would be significant digits.

In computations, figures which are not significant should be dropped from the final answer. This practice avoids implying a spuriously high degree of accuracy in a calculated result. A general rule is: Do not retain a greater number of significant digits in a result computed from multiplication and/or division than the least number of significant digits in the data from which the result is computed.

As an example, if the height and diameter of a right circular cylinder are measured to the nearest thousandth inch as 2.365 and 1.785 in., respectively, the volume could be calculated as follows:

$$V = \frac{\pi d^2 h}{4}$$

$$= \frac{3.1416 \times (1.785 \text{ in.})^2 \times 2.365 \text{ in.}}{4}$$

$$= 5.9183205 \text{ in.}^3 \text{ if eight digits are retained}$$

But this implied accuracy is not justified. Actually, we are justified in retaining only four significant digits in the answer, so it should be "rounded off" to 5.918 in.³. In dropping digits that are not significant, the last digit retained is increased by 1 if the first rejected digit is 5 or greater. For example, in the above calculation, if the seven-digit answer had been 5.9186205 in.³, the proper four-digit result would be 5.919 in.³.

SUMMARY

Laboratory work is enjoyed by most students. Indeed, if you regard the performance of laboratory investigations and the subsequent preparation of laboratory reports with distaste, it is probable that a choice of a career involving research or management in science, engineering, business, agriculture, or allied health would not be a wise one for you. Laboratory research, problem solving, and report writing constitute the daily work of the professional man or woman in technical fields, and if you enjoy these activities as a student, your future in science or technology holds a great deal of promise.

Approach every laboratory exercise with an attitude of serious inquiry; make every laboratory report a project for self-improvement; see that reports are models of neatness and organization; and finally, turn them in on time!

PART ONE ■ MEASUREMENT, STATICS, AND MECHANICS

EXPERIMENT 1

PRECISION MEASUREMENT

PURPOSE

To study some instruments and methods of precision measurement and to compute the volume and density of a metal cylinder, a steel ball, and a hardwood cone.

APPARATUS

Steel rule; vernier and micrometer calipers (English and metric scales); comparator with dial indicator accurate to 1/1000 in.; set of gage blocks; laboratory balance and a set of masses, or a direct-reading digital balance; charts or manufacturer's instructions explaining construction and methods of reading the measuring instruments; metal cylinder about 3 in. long and 1 in. in diameter; smooth steel ball with a nominal diameter of 1 in.; hardwood cone with circular base (or other geometrical forms provided by the instructor). (A worn fuel injector, or similar auto part, would provide a good exercise in measurement.)

INTRODUCTION AND BACKGROUND THEORY

Measurements are essential to modern science and industry. Mass production and interchangeability of parts would be absolutely impossible without accurate measurements and associated quality control. The student of physics must be thoroughly familiar with both the metric and English systems of measurement and with the more common measuring instruments used in industry.

In the building construction industry, measurements to 1/32 in. are usually satisfactory. Sheet-metal construction involves measurements to 0.01 in. The auto industry is tooled up for measurements accurate to 0.001 in., and the aircraft industry to 0.0001 in. Space technology is requiring measurements of 0.00001 in. accuracy. Scientific instruments may involve measurements to an accuracy of a millionth part of an inch. Medicine requires extremely accurate mass and weight measurements. See Table I of the Appendix for metric–English equivalents.

MEASUREMENTS

With the instruments at hand and with the aid of manufacturers' charts, practice until accurate readings from vernier calipers and micrometer calipers can be obtained quickly and easily. (See Fig. 1.1.) Calibrate the dial indicator by using the gage blocks.

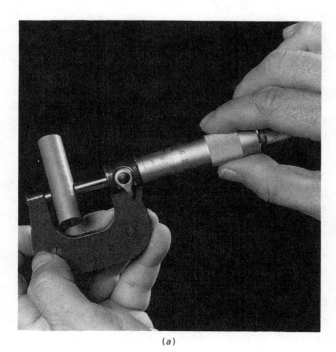

(a)

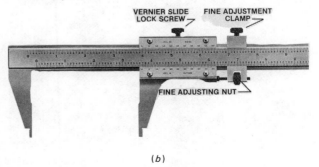

(b)

Figure 1.1. Measuring instruments. (a) Metric micrometer, in use. (b) English vernier caliper. (L. S. Starrett Co.)

Specific Steps

1. With the steel rule, take and record all measurements necessary for calculating the volume of the metal cylinder (volume $V = (\pi d^2 h)/4$) and the steel ball ($V = \pi d^3/6$). Use the suggested Data Sheet on page 5 unless your instructor requires a different format.

2. Repeat with the English vernier caliper to 0.001 in.

3. Repeat with the metric vernier caliper to 0.01 mm.

4. Check the English-system measurements on the cylinder and the ball by using the comparator with dial indicator (See Fig. 1.2). First, calibrate this instrument with a set of gage blocks. Polish the gage blocks you use with a soft cloth, to remove fingerprints.

5. Determine the masses in grams (g) of the metal cylinder and the steel ball on the beam balance (Fig. 1.3), and then check the values you obtain on a precision digital balance (if available). (See Fig. 1.3.)

6. With the steel rule and also with the metric vernier caliper (if its jaws open wide enough), measure the base diameter and the *slant height*(s) of the hardwood cone. Then determine its mass on the beam balance. *Note:* The volume of a right circular cone is given by $V = (\pi d^2 h)/12$, where h is its *vertical* height and d is the diameter of the circular base. You will have to calculate h from your measurements of the slant height and the base diameter. (See Fig. 1.4.)

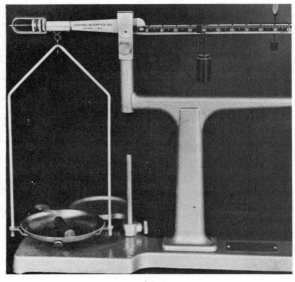

(a)

(b)

Figure 1.3. (a) A triple-beam centigram balance for determining masses. (b) Direct-reading digital balance. *(Central Scientific Co.)*

7. If an auto fuel injector (or other precision part from industry) is available, measure its key dimensions in English units with the dial indicator (previously calibrated with the gage blocks). If the manufacturer's specifications for the part are available, compare your measurements with them. Do the measurements fall within the specified tolerance?

CALCULATIONS

Compute the volume of each test object in cubic centimeters (cm^3) accurate to three significant digits. Then com-

Figure 1.2. Comparator-dial indicator set up to check length of metal cylinder. The comparator is first set and calibrated by the gage block. (Do-All Company and L.S. Starrett Co.)

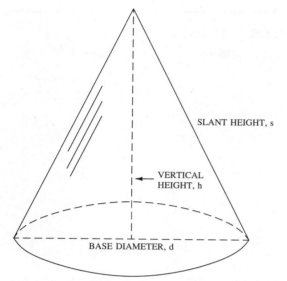

Figure 1.4. Sketch of right circular cone, showing base diameter, slant height, and vertical height.

pute the density of each, given that *density equals mass per unit volume*.

$$\rho = \frac{M}{V} \qquad (1.2)$$

Express the density of all three materials in g/cm^3 and in kg/m^3. Look up the accepted values in a handbook and calculate the percent error of your computed results for steel and hardwood.

Based on your value for the density of the metal cylinder, what metal do you think it is?

ANALYSIS AND INTERPRETATION

Review the instructions on writing laboratory reports which were given in the **Introduction.** Follow carefully any additional directions given by the instructor. Include a discussion of the following in the final section of the laboratory report:

1. With sketches, explain in detail the operation and the reading of vernier and micrometer calipers.
2. Describe how dial indicators and comparators are used in mass-production manufacturing.
3. Explain how gage blocks are used to check the accuracy of other measuring instruments in industrial laboratories and tool rooms.
4. What does *density* mean, and why is it important in the design of machine parts?
5. What are the units of density in the English system of measurement?
6. Discuss at some length the practical applications to technology and industry of the methods of precision measurement you have studied in this experiment.

NOTES, CALCULATIONS, OR SKETCHES

Name _____ Date _____ Section _____

Course _____ Instructor _____

DATA SHEET

EXPERIMENT 1 ■ PRECISION MEASUREMENT

Measurements Obtained

Object	Dimension	Measurement by				Mass (g)
		Steel rule	Vernier caliper	Micrometer caliper	Dial indicator	
Metal cylinder	Height	cm	cm		in.	g
		in.	in.		in.	
	Diameter	cm	cm	cm	in.	
		in.	in.	in.	in.	
Steel ball	Diameter	cm	cm	cm	in.	g
		in.	in.	in.	in.	
Hardwood cone	Diameter of base	cm	cm			g
		in.	in.			
	Slant height	cm	cm			
		in.	in.			
Other objects						

NOTE: This blank-form data sheet, and the ones to follow, are merely suggested formats for recording data. Get in the habit of recording data neatly and accurately as you perform experiments. The *original* data sheet, on which you record measurements *as they are actually taken,* should be a part of your laboratory report.

NOTES, CALCULATIONS, OR SKETCHES

EXPERIMENT 2

EQUILIBRIUM OF COPLANAR CONCURRENT FORCES— VECTOR ADDITION

PURPOSE

To study the equilibrium of a body under the action of coplanar concurrent forces, and to use analytical and vector methods in the solution of problems in statics.

APPARATUS

Force table, complete with centering pin, cords, ring (or disk), and pulleys; weight hangers; small builder's level; protractor and ruler; graph paper; sets of *weights* (newtons), or *masses* (kg and g).

INTRODUCTION AND BACKGROUND THEORY

A system of coplanar forces whose lines of action all pass through the same point is said to be a *concurrent force system*. Such a system of forces may be replaced by a single force through the same point, which would have the same effect or result as the original force system. This single force is called the *resultant* of the system.

Conversely, a concurrent force system can be exactly balanced by a single force. Such a balancing force is called the *equilibrant*. Its line of action is also through the point of concurrence. The *resultant* and the *equilibrant* of any concurrent system of forces are equal in magnitude and have the same line of action, *but they are oppositely directed*. Figure 2.1 shows these relationships in a *space diagram*.

In this experiment coplanar concurrent force systems will be studied with the assistance of a *force table* (see Fig. 2.2). A small metal ring (or disk) will be caused to experience a set of concurrent, horizontally-directed forces. The magnitudes and directions of these individual forces will be adjusted so that the resultant force acting on the ring will be zero, and the ring will therefore be in equilibrium.

Figure 2.2. A force table with centering pin, ring, cords, pulleys, and attached masses (the pull of gravity on which results in forces) on the cords. *(Central Scientific Co.)*

Special Note on Force and Mass Units

The forces on the metal ring of the force table will be supplied by tension in several cords attached to it and passing over pulleys (assumed frictionless) clamped to the outer rim of the force table (see Fig. 2.2). Known masses from a laboratory set are then hung from the cords, as shown. These hanging masses are acted upon by the attraction of earth's gravity to produce the forces of tension in the cords.

The preferred unit of *force* in the SI-metric system is the *newton* (N). However, if your laboratory provides sets of *masses* (in kilograms and/or grams) you will have to convert these hanging masses into forces (in newtons), as follows:

$$\text{newtons (N)} = \text{kg} \times 9.81 \quad (2.1)$$
$$= \text{g} \times 0.00981$$

These conversion factors are given at this point without derivation. Further clarification of this relationship will be provided later in Experiment 7, where it is shown that

$$\text{weight (force)} = \text{mass} \times \text{the acceleration of gravity, } g$$
$$w = mg \quad (2.1')$$

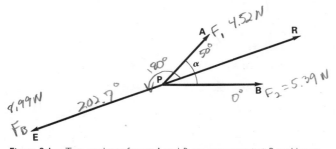

Figure 2.1. Two coplanar forces *A* and *B* are concurrent at *P*, making an angle *BPA* = α. *R* is the *resultant* of *A* and *B*. *E* is the *equilibrant*, equal in magnitude but opposite in direction to *R*.

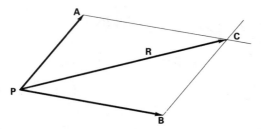

Figure 2.3. The parallelogram method of determining a resultant. **R** is the resultant of the force vectors **A** and **B**.

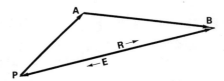

Figure 2.4. The vector triangle method of determining the resultant of two force vectors. The vector that closes the triangle represents the *resultant* if read in the direction **PB**, and represents the *equilibrant* if read in the direction **BP**. As a vector equation, **PA + AB = PB**.

Note that g for gram is not in italics, but *g* for acceleration of gravity is italic. The acceleration of gravity at the earth's surface, *g,* is 9.81 m/s². *To repeat,* for this experiment and others to follow, make use of Eq. (2.1′) to obtain *forces* in newtons from *masses* in kilograms or grams.

If *weights* in newtons are provided in your laboratory, your procedure will be straightforward. Merely observe and record these newton (N) values on the data sheet.

When the selection of directions and force magnitudes is such as to hold the ring in the center *without the presence of the centering pin,* the system is in equilibrium. In this condition any *one* of the forces can be considered as the equilibrant of the force system made up of the other forces. Displace the ring and see if it returns to the center position quickly.

Since forces have both magnitude and direction, forces are *vector* quantities. Force problems can therefore be solved by vector methods, as illustrated below.

When only two forces are involved, the resultant may be found either by the parallelogram method (see Fig. 2.3), or by the vector triangle method (Fig. 2.4).

The magnitude of the resultant of two coplanar concurrent forces may also be expressed (analytically) in terms of the two given forces and the angle between them, as follows (see Fig. 2.1):

$$R^2 = A^2 + B^2 + 2AB \cos \alpha \qquad (2.2)$$

Force vectors may be resolved into horizontal and vertical *components* from which an analytical solution of a force problem is possible. Analytically (see Fig. 2.5),

$$F_{1x} = F_1 \cos \theta_1$$

and

$$F_{1y} = F_1 \sin \theta_1$$

etc.

Let ΣF_x represent the algebraic sum of all the horizontal components of the forces in a concurrent force system, and let ΣF_y represent the algebraic sum of all the vertical components. Then

$$\Sigma F_x = F_{1x} + F_{2x} + F_{3x} + \cdots + F_{nx}$$
$$\Sigma F_y = F_{1y} + F_{2y} + F_{3y} + \cdots + F_{ny}$$

The magnitude of the resultant is

$$R = \sqrt{(\Sigma F_x)^2 + (\Sigma F_y)^2} \qquad (2.3)$$

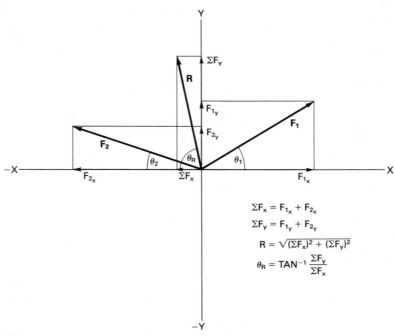

$$\Sigma F_x = F_{1x} + F_{2x}$$
$$\Sigma F_y = F_{1y} + F_{2y}$$
$$R = \sqrt{(\Sigma F_x)^2 + (\Sigma F_y)^2}$$
$$\theta_R = \text{TAN}^{-1} \frac{\Sigma F_y}{\Sigma F_x}$$

Figure 2.5. Sketch of the analytical method of solving for the resultant of a coplanar concurrent force system. Forces are first resolved into their horizontal and vertical components, and then added, with proper regard to algebraic sign.

and the angle which it makes with the horizontal (x-axis) is given by

$$\theta_R = \tan^{-1} \frac{\Sigma F_y}{\Sigma F_x} \qquad (2.4)$$

For equilibrium of a *coplanar concurrent* force system the necessary and sufficient condition is

$$\Sigma F_x = 0$$
$$\Sigma F_y = 0 \qquad (2.5)$$

MEASUREMENTS

1. Set up the force table, level it carefully, and hang two masses, in excess of 300 g each, from two of the hangers. These hanging masses result in forces. They should not be equal and should not form a right angle. They should be set up with the centering pin holding the ring in place. Now adjust a third force, as to both magnitude and direction, to bring the force system exactly into equilibrium. When this is accomplished, the third force is the *equilibrant*. Record the direction and magnitudes of the two original forces and of the equilibrant. (The resultant is equal in magnitude to the equilibrant, but is opposite in direction.)

1a. Repeat, with different values for the magnitude and direction of the two original forces.

2. Finally, set up three original forces of unequal magnitudes and not forming equal angles or right angles, and determine a fourth force which will render the system in equilibrium. Record all data.

CALCULATIONS AND RESULTS

1. Two Forces and Their Resultant and Equilibrant
First, draw *space diagrams* freehand for both trial set-ups (as in Fig. 2.1) of the actual forces as set on the force table. Label them carefully. Then solve for the *resultant* by the *parallelogram method*. This second drawing should be a large, neat drawing *to scale,* on a half sheet of graph paper whose grid is suited to decimalization.

Next, solve for the *equilibrant* by the *vector triangle method,* as in Fig. 2.4. Make an accurate vector diagram on the same sheet of graph paper. The equilibrant is the vector which closes the **vector triangle.**

Finally, solve for the resultant analytically by using Eq. (2.2).

Compare the results in each case with the actual experimental values obtained from the force table.

2. Three Forces and Their Resultant and Equilibrant
Draw the space diagram freehand as before. Then solve (on a second sheet of graph paper) by the vector polygon method for the resultant and the equilibrant. (The vector polygon method is merely an extension of the vector triangle plot. The last plotted vector should, except for experimental error, close the polygon.)

Finally, solve for the resultant (magnitude and direction) of the three forces by the *analytical method,* using the technique of resolving forces into their horizontal and vertical components (Fig. 2.5), and applying Eq. (2.3).

Compare the results with the actual experimental values obtained from the force table.

All the above calculations except the analytical solutions should be carried out on graph paper, accurately and neatly plotted. Give each plot a title, and label everything correctly. Be sure to show how the calculated results compare with the experimental values from the force table.

ANALYSIS AND INTERPRETATION

1. Explain how the experiment has illustrated the principles of vector addition. What does the vector equation $\mathbf{F}_1 + \mathbf{F}_2 = \mathbf{R}$ express? How would you write the same expression in algebraic terms?

2. If a particle or a body is in equilibrium under the action of three coplanar forces, what must be true of the lines of action of the three forces?

3. List two or three practical examples from industry in which coplanar concurrent forces in equilibrium are involved. Why is the study of forces in equilibrium (statics) so important to construction engineering?

4. If a coplanar concurrent force system is in equilibrium, what will be true of the vector diagram which represents it?

5. By means of a simple sketch show how coplanar concurrent forces are involved in bridge trusses, roof trusses, and hoisting cranes.

6. A person standing at rest on a horizontal floor is an example of a body in equilibrium while experiencing forces. What forces act *on the person*? Are the forces coplanar? Concurrent?

NOTES, CALCULATIONS, OR SKETCHES

Name Ben Botjer Date ____ Section ____

Course ____ Instructor ____

DATA SHEET

EXPERIMENT 2 ■ EQUILIBRIUM OF COPLANAR CONCURRENT FORCES—VECTOR ADDITION

	Force 1 (N)		*Force 2 (N)*		*Force 3 (N)*		*Equilibrant (N)*	
Trial	*Magnitude*	*Direction*	*Magnitude*	*Direction*	*Magnitude*	*Direction*	*Magnitude*	*Direction*
1	3.43 N	0°	1.176 N	60°			4.116 N	194.3°
1a	7.35 N	80°	.882 N	350			7.35 N	254°
2	2.45 N	10°	.686 N	60°	1.47 N	310°	3.626	178°

1) 350 ∠0° + 120 ∠60° = R = 422.9657 θ = 14.22226

420g ∠194°

2) 750 ∠80° + 90 ∠350° = R = 755.38 θ = 73.15

750g ∠254°

3) 70 ∠60° 250 ∠10° 150 ∠310° = R = 377.77 θ = -1.6

370g ∠178°

NOTES, CALCULATIONS, OR SKETCHES

EXPERIMENT 3

COPLANAR CONCURRENT FORCES—THE HOISTING CRANE PROBLEM

PURPOSE

To study the equilibrium conditions involved in the handling of materials with tackle, booms, and cranes.

APPARATUS

Laboratory crane boom with screw eye at one end and knife-edge at the other; set of hooked weights; heavy cord; V-groove clamp; heavy laboratory support rods with table clamps; pulleys and pulley clamps; large protractor; meter stick; small builder's level.

Note: Your laboratory may have crane booms with a compression spring built in, and a scale that reads directly in newtons or grams (see Fig. 3.3).

INTRODUCTION

Materials handling presents many interesting and demanding problems which require the methods of statics and the application of vectors for their solution. In actual practice, hoisting and materials-handling equipment may involve forces which are neither coplanar nor concurrent. However, if the loads being handled are assumed to be large in comparison with the weight of the cables and booms of the equipment, the assumption of concurrent forces can be made without appreciable error. In this experiment the weights of the cords and of the boom are small in comparison with the loads used, and they will be neglected, for purposes of simplification. Recall from Experiment 2 that you may use newton *weights* for loads if your laboratory supplies these. If not, and your loads are *masses*, remember

Newtons (N) = kg × 9.81
           = g × 0.00981

Two cases (or types of equipment) will be studied:
1. Fixed-boom equipment, in which the load is lifted from one location and is set down in another by means of cables controlled (in the actual industrial situation) by power winches. The cables run over pulleys (called "sheaves" in industry) at the ends of the fixed booms (see Fig. 3.1).
2. Movable-boom equipment, in which the boom can be swung around, raised, and lowered to pick up equipment and to set it down where desired.

The analysis of stresses in each case makes use of the technique of "isolating" a point at which forces are concurrent, and applying vector methods to the solution of the problem *at that point,* neglecting for the moment any con-

sideration of forces acting at other points. It is important in every case to draw first a *space diagram* of the forces acting at the point isolated and then to construct a *vector diagram* to solve for the unknown stresses of tension or compression.

MEASUREMENTS

Case I. Fixed-Boom Equipment

Set up the situation as in Fig. 3.1. The load w should be of the order of 20 N (use a 2-kg mass). P_1 and P_2 are pulleys on laboratory stands, which in an actual situation would be located on the ends of fixed booms (not shown). F_1 and F_2 are forces (weights) which bring point C into equilibrium. A small metal ring may be used for point C. Adjust the values of F_1 and F_2 until CP_2 is horizontal and angle F_1P_1C is of the order of 30 to 40°. Use small weights as necessary in order to bring C into exact equilibrium, with CP_2 horizontal. Displace the system by hand, and see if it returns to the desired equilibrium position. Point C can be considered a particle in equilibrium under the action of three forces, all of which are tensional forces in the cords. Record the exact values of w, F_1, F_2, and the angle F_1P_1C.

Repeat the observation using a load w, of 40 N and making equilibrium for point C occur when angle F_2P_2C is 60°.

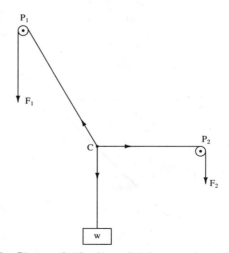

Figure 3.1. Diagram of a "fixed boom" rig for materials handling. Pulleys P_1 and P_2 are mounted in fixed positions on the ends of booms (not shown). By varying forces F_1 and F_2 (using power-driven cable winches) the load w can be lifted and landed at different positions on a dock or the deck of a ship.

Case 2. Movable-Boom Equipment

Set up the apparatus as in Fig. 3.2a and b. P_1C (the *boom stay*) should be horizontal for the first trial. Use a load of about 30 N for w. Place the knife-edge of the boom in the special V-groove clamp and adjust the force (weight) F_1 so that the boom makes an angle of about 40° (angle *CAM*) with the mast. The boom AC is now in compression. Now place pulley P_2 on a heavy support in such a position that

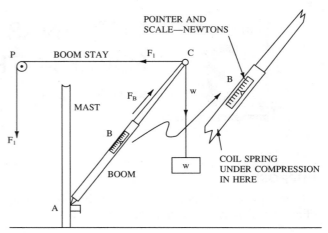

Figure 3.3. Sketch of a laboratory crane boom with built-in compression spring for direct reading of the compressional force F_B in the boom. F_B acts upward against C, as indicated. The reading at B should be in newtons.

(a)

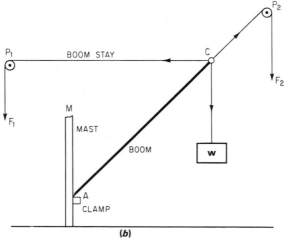

(b)

Figure 3.2. (a) Apparatus setup for studying the principles of a crane hoist. (b) Diagram of the forces acting at the end of a crane boom.

cord P_2C will be in the *exact line of the boom*. Determine the force (weight) F_2 which will just pull the boom knife-edge out of the V-groove clamp at A. This force (tension away from C) is equal to the compressional force in the boom which acts toward C. Record the exact values of w, F_1, F_2, and angle *CAM*. If you are using a boom with a built-in compression spring, read F_2 from the scale on the boom. Pulley P_2 will not be needed. (See Fig. 3.3.)

Repeat for the case in which the boom stay is not horizontal but with angle *CAM* still at about 40°. Raise the pulley P_1 until angle CP_1F_1 (the angle between the boom stay and the vertical) is about 70°. Make w about 40 N. Record all necessary data for this case.

CALCULATIONS AND RESULTS—VECTOR DIAGRAMS

(All diagrams are to be made on graph paper.)

Case 1 For both trials, draw the *space diagram* for the forces acting (concurrently) at C. Show by arrowheads that they all act away from C (tension). This process is known as *isolating* point C.

Now plot the *vector diagram* (Fig. 3.4) starting with a vector representing w, since both its magnitude and its direction are known. Make the diagram large enough for good accuracy. Next (heel-to-toe) draw a line for F_2 whose direction is that of CP_2 (see Fig. 3.1) but whose magnitude is unknown, except as you have determined it experimentally. Since, when three coplanar forces cause equilibrium, they must be concurrent, the third vector must close the vector polygon. To draw the vector representing F_1 start at the "feathered" end of the w vector, and draw a line whose direction is that of the CP_1 force. The intersection of the CP_1 line with the CP_2 line establishes the vector triangle.

From the vector triangle just plotted determine the magnitudes of F_1 and F_2. Compare these values with the actual values recorded from the experiment. Calculate the percent differences.

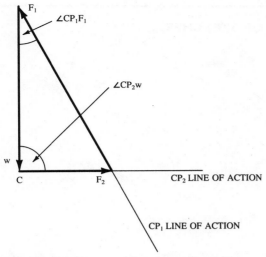

Figure 3.4. Example of a vector diagram for Case 1. Your diagram must be drawn to exact scale on graph paper.

Case 2 For both trials, isolate point *C* and draw the space diagrams. Then draw the vector diagrams in accordance with the method explained above. Determine the force of compression in the boom, and the force of tension in the boom stay, and compare these values with the experimental results for both trials.

ANALYSIS AND INTERPRETATION

Explain the sources of error (difference between experimental results and results from the vector treatment) in this experiment. Why are the differences greater in Case 2 than in Case 1? (Recall that the weight of the boom was assumed negligible.)

Why is the technique of "isolating" a point of such importance in the solution of problems in statics?

Explain a materials-handling situation in which the Case 1 equipment might be used. The Case 2 equipment.

What assumption is inherent in the statement that "a cable or rope can sustain only tensional stresses"?

From collateral reading explain the meaning of the following terms, as applied to materials-handling equipment:

1. Block and tackle
2. Running rigging and standing rigging
3. Sheaves
4. Yard and stay rig
5. Winch
6. Pallet
7. "Take a strain"
8. "Lower the boom"
9. Forklift
10. Motorcrane

Name ______________________ Date ____________ Section ____________

Course ______________________ Instructor ______________________

DATA SHEET

EXPERIMENT 3 ■ COPLANAR CONCURRENT FORCES—THE HOISTING CRANE PROBLEM

Case 1. Fixed Boom

Trial	*Load w (N)*	F_1 (N)	F_2 (N)	*Angle* F_1P_1C *(degrees)*	*Angle* F_2P_2C *(degrees)*
1					90°
2					

Case 2. Movable Boom

Trial	*Load w (N)*	*Angle CAM (degrees)*	F_1 (N)	F_2 (N)	*Angle* CP_1F_1 *(degrees)*
1					90°
2					

NOTE: The load and force units suggested above (N) may be changed to accord with the instructor's wishes and the equipment available. The pound (lb) could just as well be used.

NOTES, CALCULATIONS, OR SKETCHES

EQUILIBRIUM OF COPLANAR PARALLEL FORCES—THE PRINCIPLE OF MOMENTS

PURPOSE

To determine the conditions for equilibrium of a rigid body under the action of a system of *coplanar parallel forces;* to study the principle of moments and the loading of a simulated beam.

APPARATUS

Meter stick and supports; sliding clamps with knife-edges; set of hooked masses; triple-beam or other suitable balance; pulleys; cord; laboratory stands; unknown mass (of the order of 2 to 4 kg). (See Fig. 4.1.)

INTRODUCTION

In the previous experiments (coplanar concurrent forces), since the lines of action of the forces all passed through the same point, there was no turning effect on the body. Any unbalanced force would merely cause a linear acceleration. However, if a rigid body is acted upon by a system of forces which are *not concurrent*, there may result either a linear acceleration or an angular acceleration (or both) unless the magnitudes, lines of action, and points of application of the forces are so chosen as to produce equilibrium. In this experiment the conditions for equilibrium of a *coplanar nonconcurrent* force system will be studied. The specific case of *coplanar parallel* forces will be selected.

Moment   The turning effect of a force is called *moment*. The word *torque* is also used in this connection. The moment of a force is defined as the product of the force

acting on a body and the *perpendicular distance* from the *line of action* of the force to the axis of rotation of the body. This perpendicular distance from the action line of the force, to the center for possible rotation of the body is called the *moment arm* of the force. (See Fig. 4.2.) A moment is said to be *clockwise* (considered negative) if its effect would be to rotate the body clockwise, and counterclockwise (positive) if its effect is to rotate the body counterclockwise.

Rigid Body   A rigid body is one which will transmit a force undiminished throughout its mass. The particles of a rigid body do not change positions with respect to one another. A rigid body is in equilibrium when both its linear acceleration and its angular acceleration are zero. The two conditions for equilibrium of a coplanar force system may be stated as follows:

First condition: The vector sum of all forces acting on the body must be zero. Mathematically,

$$\Sigma F = 0 \qquad (4.1)$$

Second condition: The algebraic sum of all moments about any axis (within or outside the body) must be zero.

$$\Sigma M_p = 0 \qquad (4.2)$$

where *P* may be *any point in the plane of the forces*, whether inside or outside the rigid body.

The rigid body for this experiment is a laboratory meter stick. Of course, it is not completely rigid, but it is as-

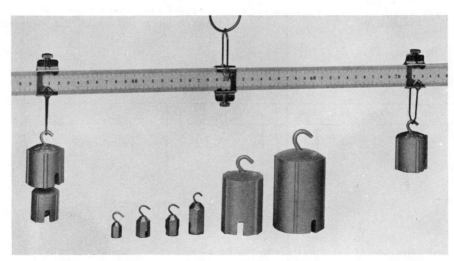

Figure 4.1.   Equipment used for studying the equilibrium of a rigid body under the action of coplanar forces. *(Welch Scientific Co.)*

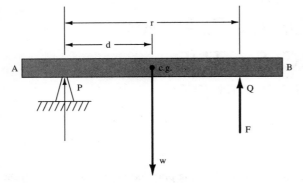

Figure 4.2. Moment and moment arm. Let AB be a beam resting on a support at P. Two other forces, F upward, and w downward, act on the beam as shown. Taking P as the fulcrum, F has a moment arm of r and moment +Fr (counterclockwise) around P. In like manner, the moment arm of w is d, and its moment around P is −wd (clockwise). There is an upward force at P, but it has zero moment since its line of action passes through P with a moment arm of zero.

sumed to be so for this exercise. The forces acting on it will be supplied by hooked masses and their clamp/supports[1]; by the mass of the meter stick itself; and by the support system (see Fig. 4.1).

As an example of the conditions for equilibrium, refer to the diagram of Fig. 4.3.

To satisfy Condition 1 (considering upward forces positive and downward forces negative), the following must be true to prevent *linear acceleration:*

$$\Sigma F = F_2 - F_1 - w - F_3 = 0 \qquad (4.3)$$

where w is the weight of the meter stick, assumed to act at the center of gravity, c.g.

To satisfy Condition 2 (no angular acceleration), the moment equation for moments taken around P (for example) is

$$\Sigma M_p = F_1 d_1 + w d_w + F_3 d_3 - F_2 d_2 = 0 \qquad (4.4)$$

Any point in the plane of the forces could have been chosen for point P, whether inside or outside the body.

---

[1] If your laboratory provides sets of *weights* in newtons, observe and record all values directly. If *masses* (kg and g) are used, remember that $w = mg$, and $N = kg \times 9.81 = g \times 0.00981$ (see Experiment 2—Special Note).

MEASUREMENTS

Determine the mass of the meter stick to an accuracy of 0.01 g. Then calculate its *weight* (newtons) from $w = mg$. In like manner obtain weights of the meter-stick clamps and keep track of the weight of each clamp, separately. Locate the center of gravity of the meter stick by balancing it with a knife-edge on the fulcrum stand. Record all these data.

Part I. Verifying First Condition for Equilibrium

Place a clamp on the meter stick at the 70-cm mark and put the knife-edge of this clamp in the fulcrum stand. Tie a strong cord to the top of this clamp, and lead it *vertically* up and over a pulley which is held in place by a laboratory stand.

On a clamp at the 95-cm mark place a load of 1 kg. At the 40-cm mark attach a clamp and a 500-g load. Finally, add masses to a clamp at the 20-cm mark until equilibrium is attained. Calculate all forces (weights) from $w = mg$, including the weight of each clamp and its exact location.

Now, with the entire system in equilibrium, add masses to the cord running up over the pulley until the knife-edge is barely lifted off the fulcrum. Calculate and record this force (weight). This force is equal to the upward force being exerted by the fulcrum on the meter stick. It is the *equilibrant* of the parallel system of forces acting downward on the rigid body (meter stick).

Part 2. Verifying Second Condition for Equilibrium

Set the clamp which will serve as the fulcrum at the 60-cm mark. Attach an unknown mass (between 0.50 kg and 1.1 kg) at the 90-cm mark. Put a 200-g mass at the 45-cm mark. Then using a 500-g mass, slide its clamp along the meter stick until equilibrium is attained. Compute the magnitudes of the forces acting on the meter stick due to the weights of the 200-g mass, the 500-g mass, the unknown mass, and the weight of the meter stick. Record all the forces and the locations at which they act.

Determine the value of the unknown mass with a laboratory balance and record the value on the data sheet.

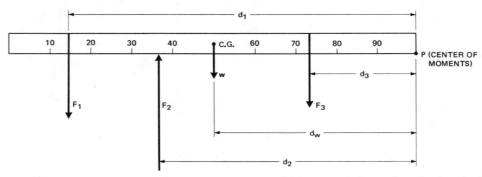

Figure 4.3. Sketch of forces involved in the equilibrium of a rigid body (coplanar, parallel force system). A general case for discussion. This sketch does not represent any of the laboratory setups.

Part 3. Loading a Beam

Support one end of a meter stick at the 5-cm mark and the other at the 80-cm mark. Thus the meter stick will simulate a beam and the laboratory supports will act as supporting columns or walls (see Fig. 4.4). From clamps at 35 cm, 65 cm, and 95 cm, hang loads of 1 kg, 300 g, and 500 g, respectively.

Directly above the left (L) and right (R) supports, rig pulleys on laboratory stands. Attach cords to the meter stick clamps at L and R and run them over the pulleys and attach mass hangers. Be sure the cords pull *vertically* upward on the meter-stick hangers. Add masses slowly to each pulley-supported mass hanger until the loaded meter stick *just lifts off* the L and R supports. Adjust everything carefully, taking off small masses and replacing them as needed until you are sure the system is in equilibrium, with the meter stick *horizontal*. Compute the weights of the 1-kg, 300-g, and 500-g masses (F_1, F_3, F_4) and their positions, and also the weights $w_1 = R_L$ and $w_2 = R_R$. Remember to include the weights of any clamps used. F_2 is, of course, the weight of the meter stick. Check the exact position of the meter stick's center of gravity. It may not be precisely at the 50-cm mark.

CALCULATIONS AND RESULTS

Draw careful space diagrams (fully labelled) of all the experimental setups.

Part 1   From the data taken in this section, use Eq. (4.1) and verify the first condition of equilibrium. Be sure to display your results in such a way that your proof will be clear. Show your force equation clearly.

Part 2   From the data taken in this section, and using the second condition, Eq. (4.2), compute the force contrib-

uted by the unknown mass. Use the fulcrum at the 60-cm mark as the center of moments. Compare this calculated result with the actual weight of the unknown object. Do your results verify the principle of moments? Show your force and moment equations.

Part 3   The reactions R_L and R_R were *experimentally* determined in this part of the experiment. Making use of the *two conditions for equilibrium* (Eqs. 4.1 and 4.2), *calculate* the values of R_L and R_R. Remember the weight of the meter stick itself and the weights of any clamps.

Compare your experimental results with the calculated results and express the percent error, accepting the calculated results as being correct. Show all your equations clearly.

ANALYSIS AND INTERPRETATION

Discuss the sources of error in the experiment. List and give a brief discussion of at least three structural components or machine components which are examples of a system of coplanar parallel forces in equilibrium.

Answer the following:

1. What is meant by a rigid body?
2. What is the definition of the moment of a force?
3. State the two conditions for equilibrium of a rigid body acted upon by a system of coplanar parallel forces.
4. In Part 2 what convenience was secured by taking the fulcrum as the center of moments?
5. Actually, what is meant by the "center of gravity" of an object?
6. A large beam is to be supported by columns of steel at either end. The beam will support a floor on which heavy machinery is to be permanently installed. Is it necessary that the steel support columns be of equal strength? Explain.

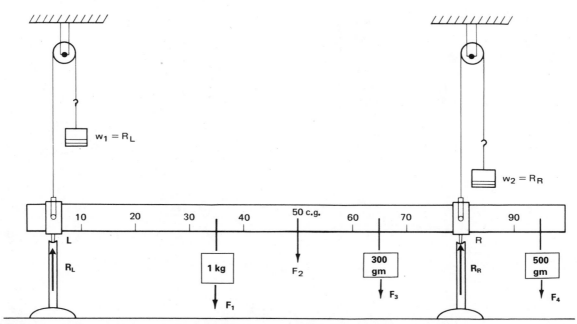

Figure 4.4.   Meter stick simulating a loaded beam acted upon by a system of coplanar parallel forces.

NOTES, CALCULATIONS, OR SKETCHES

Name _____ Date _____ Section _____

Course _____ Instructor _____

DATA SHEET

EXPERIMENT 4 ■ EQUILIBRIUM OF COPLANAR PARALLEL FORCES— THE PRINCIPLE OF MOMENTS

General Data

Mass of meter stick _____ g

Location of center of gravity of meter stick _____ cm

Masses of meter stick clamps

No. 1 _____ g   No. 2 _____ g   No. 3 _____ g

No. 4 _____ g

Data for Part 1

Fulcrum stand at _____ cm.   Meter stick weight = _____ N.

At _____ -cm mark:

Clamp mass _____ kg + load _____ kg = _____ kg. (total)

Force at _____ -cm mark = _____ N.

At _____ -cm mark:

Clamp mass _____ kg + load _____ kg = _____ kg. (total)

Force at _____ -cm mark = _____ N.

At _____ -cm mark:

Clamp mass _____ kg + load _____ kg = _____ kg. (total)

Force at _____ -cm mark = _____ N.

Data for Part 2

Fulcrum stand (center of moments) located at _____ cm.

Unknown weight w located at _____ cm.

For force F_1 at _____ -cm mark:

Clamp mass _____ kg + load _____ kg = _____ kg. (total)

Force at _____ -cm mark = _____ N.

Location of 500-g weight F_2 to bring about equilibrium _____ cm.

For unknown weight:

Clamp mass _____ kg + load _____ kg = _____ kg. (total)

Unknown weight = _____ N.

Meter stick weight = _____ N.

Data for Part 3

For load F_1 at _____ cm:

Clamp mass _____ kg + load _____ kg = _____ kg. (total)

Force F_1 at _____ -cm mark = _____ N.

Load F_2 = weight of meter stick = _____ N.

For load F_3 at _____ cm:

    Clamp mass _____ kg + load _____ kg = _____ kg. (total)

    Force at _____ -cm mark = _____ N.

For load F_4 at _____ cm:

    Clamp mass _____ kg + load _____ kg = _____ kg. (total)

    Force at _____ -cm mark = _____ N.

For upward force at left support at 5-cm mark:

    Mass at end of cord = _____ kg. $w_1 = R_L =$ _____ N.

For downward force at right support at 5-cm mark:

    Mass at end of cord = _____ kg. $w_2 = R_R =$ _____ N.

ACCELERATION OF GRAVITY—
FALLING BODIES

PURPOSE

To study free fall and to determine the acceleration of gravity g in the laboratory.

APPARATUS

Three methods are described. They should all give results within 3 to 5 percent of the accepted value for g. If apparatus is available, it is suggested that two of the three methods described be used.

Method 1 Central Scientific Co.-Behr free-fall apparatus with spark generator; synchronous spark timer and coil with high-voltage output; sensitized paper tape.

Method 2 Pasco Scientific digital free-fall apparatus with steel balls.

Method 3 Polaroid camera (Colorpak II, or equal); Strobotac (General Radio 1531-AB, or Pasco Scientific SF-9211, or equal); shiny steel ball; 2 meter sticks with large, black scale markings.

INTRODUCTION

Aristotle and other ancient philosophers reasoned (mistakenly) that the heavier a body is, the faster it should fall. The Dutch mathematician Simon Stevinus (1548–1620) was the first to show experimentally that all bodies, regardless of their masses, experience the same acceleration as they fall freely toward the earth under the influence of the force of gravity. (Such experiments are also often attributed to Galileo, conducted, according to legend, from the Leaning Tower of Pisa.) Strictly speaking, free fall experiments must be conducted in a vacuum so that the force of air resistance does not affect the results. For relatively small, smooth bodies of considerable density, however, the error introduced by conducting such experiments in the atmosphere is quite small.

In any motion problem it should be apparent that three variables—*distance*, *rate*, and *time*—are involved. If the motion is uniform, or if the concept of *average velocity* is used for motion in one direction, the motion can be described by the simple equation

$$s = \bar{v}t \tag{5.1}$$

where s = distance traveled in time t and $\bar{v}$ = *average* velocity for the time interval t.

When motion is nonuniform, that is, where velocity is

changing, *acceleration* is said to take place. If the acceleration is uniform, as from a constant force such as the force of gravity, the acceleration can be defined as the average *rate of change of velocity*, and it is given by the following equation:

$$a = \frac{v_2 - v_1}{t} \tag{5.2}$$

where $v_2 - v_1$ represents the change in velocity which occurs in time t.

If a body starts from rest (i.e., $v = 0$) and is uniformly accelerated by a constant force for a time interval t, the total distance it will travel is given by the equation

$$s = \tfrac{1}{2}at^2 \tag{5.3}$$

For the case of a body falling freely from a height h under the influence of the gravitational force, the acceleration a will be equal to the acceleration of gravity g, and Eq. (5.3) becomes

$$h = \tfrac{1}{2}gt^2 \tag{5.4}$$

Any of the following methods may be used to study falling bodies and to determine the value of g.

MEASUREMENTS

Method I. Behr Free-Fall Apparatus

Level the apparatus (see Fig. 5.1) by means of the base leveling screws and a plum bob so that the upright bar will be truly vertical. Connect the electromagnet at the top of the apparatus to a source of 6 volts (V) direct current (dc) through a switch and a rheostat of about 700 ohms (Ω). Make the necessary connections and adjustments to the spark generator so that sparks of a known frequency will jump the gap between the plummet and the apparatus as the plummet falls.

The plummet must fall freely, not touching the body of the apparatus as it falls. Cushion its impact with some sponge rubber, felt, or sand in the bottom "well." Test the fall and the spark action several times prior to putting the sensitized paper tape in place along the entire vertical surface of the Behr apparatus. The plummet must be absolutely stationary when suspended from the electromagnet. Drop the plummet by using a variable rheostat in the dc circuit rather than by merely opening the switch. (Why?)

The time interval between successive sparks is known

Figure 5.1. Behr free-fall apparatus for the laboratory study of the acceleration of a freely falling object. *(Central Scientific Co.)*

distance from hole zero to hole one. Call that distance $s_{0,1}$. The distance from hole zero to hole two is $s_{0,2}$, and so on. Record these as a *table of distances* on the Data Sheet.

CALCULATIONS AND RESULTS—METHOD I

The average velocity of an object moving in one direction in a straight line is obtained from the relation:

$$v_{av} = \frac{\text{distance}}{\text{time}} = \frac{s}{t}$$

Let $s_{0,2}$ be the distance from point 0 to point 2, and let $s_{1,3}$ be the distance from point 1 to point 3, etc. Note that $s_{1,3} = s_{0,3} - s_{0,1}$; $s_{2,4} = s_{0,4} - s_{0,2}$; etc. Let the average velocity of the falling plummet from point 0 to point 2 be v_1. Then

$$v_1 = \frac{s_{0,2}}{2t}, \; v_2 = \frac{s_{1,3}}{2t}, \; \ldots, \; v_{14} = \frac{s_{13,15}}{2t}$$

and so on. Prepare a *table of velocities* from these calculations.

Acceleration is defined as change in velocity per unit time and is given by Eq. (5.2).

Split your table of velocities into two parts and prepare

from the frequency of the spark timer. In the following discussion it is assumed to be 1/60 second(s).

After you are satisfied that all components are working satisfactorily, put the sensitized paper tape on the vertical upright of the apparatus, stretching it smoothly with the sensitized surface out. Put the plummet on the energized electromagnet, and steady it in place. Start the spark timer, and then release the plummet for its fall. The paper tape with spark holes burned in it is your record of the location of the falling body at different points in time as it fell.

Suppose the sketch of Fig. 5.2 represents a section near the beginning of the sensitized tape with the spark-hole record of the falling body. Tear off the length of tape that contains your spark-hole record and smooth it out on a flat surface. Start with the *first good, clear spark hole*, and label it zero (0). Take *every other one*, labeling them 1, 2, 3, 4, etc. until you have about 16 numbered holes. If you number every other (i.e., alternate) hole, the time interval between numbered holes will be

$$t = \frac{1}{30} \text{ second}$$

Place a meter stick (or better, a two-meter stick) on the tape, on edge (to eliminate parallax problems) and measure the distance from hole zero to each of the other numbered holes. The shortest of these distances will be the

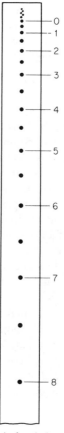

Figure 5.2. Suggested method of analyzing spark holes on the sensitized tape for the Behr method. Numbering every other hole reduces the number of measurements and the number of calculations without materially reducing the precision of the results.

a table of *accelerations* as follows (assuming 16 numbered holes, including hole number zero):

$$a_1 = \frac{v_8 - v_1}{7t}$$

$$a_2 = \frac{v_9 - v_2}{7t}$$

a_3, etc.

$$a_7 = \frac{v_{14} - v_7}{7t}$$

Average these 14 or 15 values to obtain your final result for the acceleration of gravity in cm/s^2. Convert to ft/s^2. Compare your answers with the accepted values for *your latitude and elevation above sea level*. At 45° latitude, at sea level,

$$g = 980.6 \text{ cm/s}^2 = 32.17 \text{ ft/s}^2$$

Method 2. Pasco Scientific Digital Free-Fall Apparatus

The Pasco digital free-fall apparatus (Fig. 5.3) consists of a spring-loaded ball release mechanism, a ball receptor pad, and an electronic digital timer to display the time required for a steel ball to fall distances as great as two meters.

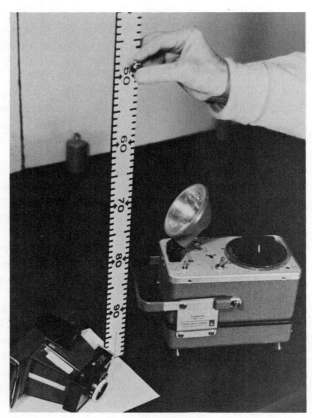

Figure 5.4. Polaroid camera and "Strobotac" unit (General Radio 1531-AB) for obtaining a photographic record of the falling steel ball.

Set up the apparatus so that the distance the steel ball falls, h, is two meters. Obtain five experimental values for the time required for the ball to drop two meters. Use Eq. (5.4) and the average value for the five time-of-fall measurements to calculate an experimental value for g. Repeat, using four other values of h, each considerably different from the first value.

CALCULATIONS AND RESULTS—METHOD 2

Use the values g obtained from time-of-fall data for the five different heights to calculate an average value for g. Calculate the percent error of your final result from the accepted value of g for your latitude and altitude above sea level. (*Optional:* Repeat the above steps using a steel ball with a different mass to demonstrate that g does not depend on the mass of the ball.)

Method 3. Polaroid Camera–"Strobotac" Method

This method consists of obtaining a photographic record of a falling steel ball by means of reflected light supplied in timed flashes from a stroboscopic lamp (see Fig. 5.4).

Mount two meter sticks end-to-end vertically, against a black cloth (or paper) background, so that a free fall of 2 m can be observed. Be sure that this 2 m scale is truly *vertical*. Check the film in the Polaroid camera and position the camera tripod at a distance that will allow the entire 2-m height to be in the field of view when the cam-

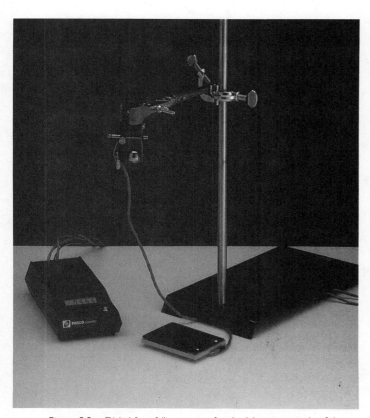

Figure 5.3. Digital free-fall apparatus for the laboratory study of the acceleration of a freely falling object. *(Pasco Scientific)*

era is focused sharply on the meter-stick scale readings. Place the Strobotac unit so that its flashes will illuminate the full length of the meter-stick scales. Provide a cushioned receptacle at the foot of the meter stick to catch the falling ball.

Now dim the room lights and turn on the Strobotac. Drop the ball a few times and make any adjustments required to assure that the ball falls freely and *very close to the edge* of the meter-stick scales. Observe the pattern of reflected flashes visually, as the ball falls.

To obtain a trial for the record, with the Strobotac flashing, one person should open the camera shutter just before a second person drops the steel ball. Close the camera shutter and develop the print in accordance with the Polaroid camera instructions.

Repeat for two more drops, or until you have three clear prints of the sequential positions of the falling ball over the 2-m distance. Note and record the flash frequency of the Strobotac unit.

With a magnifying glass, determine the successive locations of the falling ball as referenced to the meter-stick scale. Make a table of distances, then a table of velocities, and finally a table of accelerations, as explained under Method 1, above. Determine a mean value of g from the three trials. Compare it with the accepted value for your latitude and elevation above sea level.

ANALYSIS AND INTERPRETATION

(The same for any method)

1. Analyze and discuss the sources of error inherent in the method of performing the experiment.

2. Based on collateral reading, describe Stevinus's (or Galileo's) experiments by which it was proved that Aristotle and the ancient philosophers were mistaken about falling bodies.

3. Explain, from a purely reasonable analysis, why acceleration imparted to a body by gravity is independent of the mass of the body.

4. How does g vary with latitude and altitude above sea level? Why?

5. It is estimated that a good high jumper (7.5 ft on earth) could ''go over the bar at about 40 ft'' on the moon. Why?

6. Discuss some of the problems which the force of gravity poses for aerospace and satellite technology—on earth and on other celestial bodies.

7. Suppose that by some suitable means the falling object had been given a downward push just at the instant of release, rather than just being dropped. Would the value of g obtained in this experiment have been increased, decreased, or unaffected by this push? Explain your answer. (*Hint:* Note that according to Eq. (5.2) the acceleration depends on how rapidly the velocity *changes*, but not on how fast the falling object is moving.)

Name ______________________ Date ____________ Section ____________

Course ______________________ Instructor ______________________

DATA SHEET

EXPERIMENT 5 ■ ACCELERATION OF GRAVITY—FALLING BODIES

1. Behr Free-Fall Method

Spark timer interval ____________ second

Table of Distances

$s_{0,1}$ ____________

$s_{0,2}$ ____________

$s_{0,3}$ ____________

$s_{0,4}$ ____________

$s_{0,5}$ ____________

$s_{0,6}$ ____________

$s_{0,7}$ ____________

$s_{0,8}$ ____________

$s_{0,9}$ ____________

$s_{0,10}$ ____________

$s_{0,11}$ ____________

$s_{0,12}$ ____________

$s_{0,13}$ ____________

$s_{0,14}$ ____________

$s_{0,15}$ ____________

Latitude at your location ____________

Elevation above sea level ____________

2. Pasco Scientific Method

	Digital timer readings					
Height of drop h	1	2	3	4	5	*Mean time of fall t*

Latitude ____________

Elevation above sea level ____________

3. Polaroid–"Strobotac" Method

Design your own data sheet using an analysis similar to that explained in Method 1, above.

EXPERIMENT 6

PROJECTILE MOTION

PURPOSE

To study the flight of a projectile in the laboratory and to determine the initial velocity of a projectile from measurements of range, height of fall, and/or angle of elevation of a gun.

APPARATUS

Spring gun with metal ball projectile; plumb bob; large protractor or inclined plane with protractor attached; heavy corrugated paperboard to cover impact area (or a catch box); cardboard sighting tube; carbon paper; masking tape; meter stick or steel measuring tape.

INTRODUCTION

A *projectile* is defined as any object in motion through space or through the atmosphere which no longer has a force propelling it. Thrown balls, rifle bullets, and falling bombs are examples of projectiles. Rockets and guided missiles are not projectiles while the propellant is burning, but become projectiles once the propelling force, guidance forces, or both, cease to exist.

A projectile has no on-board propelling force nor guidance system. This laboratory study neglects all forces acting on the projectile except the force of earth's gravity.

We shall study two cases of projectile motion:

Case 1. Horizontal Projection

In Fig. 6.1, a projectile is discharged horizontally with an initial horizontal velocity v. It is immediately acted upon by the force of gravity and is accelerated vertically downward, all the while retaining its initial horizontal velocity (air resistance neglected). Its *trajectory* or flight path is shown by the curved line *PMX*.

If the projectile is discharged horizontally from a point directly over point G from a height h, it will fall on target X, whose theoretical horizontal range R is given by the equation

$$R = v\sqrt{\frac{2h}{g}} \tag{6.1}$$

Both R and h can be measured, and if air resistance is assumed zero, the initial horizontal velocity of the projectile can be obtained by solving Eq. (6.1) for v:

$$v = R\sqrt{\frac{g}{2h}} \tag{6.2}$$

In the case of an aircraft in horizontal flight, dropping a bomb or other object, R is of course unknown. The *line of sight* (LOS) to the target is of special interest in this case since it can be accurately determined by suitable instruments. Let θ be the angle which the LOS makes with the horizontal at the instant when the bomb should be released in order to hit the target X. The angle θ is known as the *angle of depression*. From the geometry of Fig. 6.1

$$tan\ \theta = \frac{h}{R} = \frac{h}{v\sqrt{2h/g}}$$

whence, simplifying,

$$tan\ \theta = \frac{1}{v}\sqrt{\frac{gh}{2}} \tag{6.3}$$

which yields the correct value of θ for the known values of the altitude and velocity of the launching aircraft.

Case 2. Projection at an Angle Above the Horizontal

Fig. 6.2 represents in diagram form the simplest case of the "baseball player's problem." The same analysis also applies to surface gunnery when gun and target are in the same horizontal plane. The projectile leaves the outfielder's fingers, or the gun muzzle, at P with an initial velocity v at an angle of elevation θ. As soon as it leaves, it comes under the influence of the force of gravity and is accelerated downward, falling below the LOS. Its path is a parabola (neglecting air resistance), and it lands at X on the same horizontal level as P, having covered a horizontal range R. The maximum height to which it rises is h_{max}. The time of rise is equal to the time of fall; the horizontal component of the velocity v_x is a constant; (horizontal motion and vertical motion occur simultaneously, one not

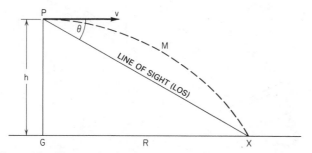

Figure 6.1. Diagram showing the relationships involved in the horizontal launch of a projectile.

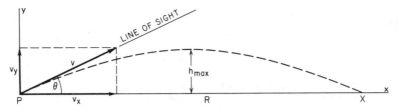

Figure 6.2. The baseball player's (or surface gunnery) problem in projectile motion. *X* and *P* are in the same horizontal plane.

affecting the other); and the vertical component v_y has the same *magnitude* at *X* as it does at *P* (air resistance neglected).

The theoretical range of a projectile launched or fired under these conditions is

$$R = \frac{v^2 \sin 2\theta}{g} \qquad (6.4)$$

from which

$$v = \sqrt{\frac{Rg}{\sin 2\theta}} \qquad (6.5)$$

The theoretical maximum height to which the projectile will rise is

$$h_{\max} = \frac{v^2 \sin^2 \theta}{2g} \qquad (6.6)$$

An outfielder "rifling" a throw to home plate to cut off a run must solve this problem instinctively. This case applies also to golf, tennis, and other games where a ball is thrown, batted, or driven at an angle above the horizontal.

MEASUREMENTS

1. Horizontal Projection (Fig. 6.1.) Set the spring gun (Fig. 6.3) for horizontal firing from a height *h*, about 3 or

4 ft above the floor. Place a square of corrugated paperboard to protect the floor from the ball's impact, or better still, provide a "catch box" (see Fig. 6.4) to prevent damage to the floor and to prevent the ball from rolling around the laboratory. The exact point of impact of the ball can be recorded by allowing it to land on a piece of carbon paper which has white paper underneath. A second piece of paper on top of the carbon paper will prevent its being torn by the impact. Secure the target papers in the catch box at a location that covers point X.

Fire three shots in rapid succession without any change in the setup, and determine the mean point of impact (MPI). Measure *R* and *h* as precisely as possible. The point *G* (Fig. 6.1) on the floor is to be determined with a plumb bob. Sight from *P* through the cardboard tube at the MPI (on the floor), and with a large protractor determine the value of angle θ as accurately as the protractor will allow. The angle θ may also be calculated from the measured values of *R* and *h*.

Repeat for two other values of *h*. Record all data on the Data Sheet.

2. Upward Projection. The Baseball Player's or Surface Gunnery Problem Clamp the spring gun to an inclined plane, and elevate the plane to a 20° angle. (Some models of the spring gun have their own elevating mechanism.) Arrange the gun so that the projectile will land on a surface *whose horizontal plane is the same as that which passes through the ball as it leaves the gun* (Fig. 6.2).

Fire three shots as before and determine the MPI. Measure *R*. By sighting across the ball *while it is in flight*, get a rough approximation of $y_{\max}(= h_{\max})$ by averaging the observations from the three trials.

Repeat for two other values of θ, one near 30°, and one about 40°. Record all the data.

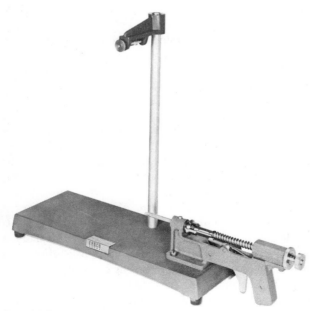

Figure 6.3. Spring gun used for firing the projectile ball. (*Central Scientific Co.*)

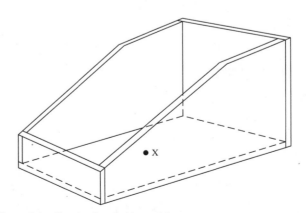

Figure 6.4. Sketch of a suitable catch box.

CALCULATIONS

Using Eqs. (6.2) and (6.5), calculate the "muzzle velocity" of the projectile from the spring gun for the six trials, and average the results.

Compare the observed values of θ (Case 1) with values calculated from measured values of h and R.

Compare the values of h_{max} observed in Case 2 with theoretical values calculated from Eq. (6.6).

Select a value of θ (Case 2) and a value of R somewhat less than R_{max}, and calculate the height h for these selected values. Make a 3-in. diameter hole in a piece of cardboard, and hold it at your predicted R, h position. Fire the gun. Does the ball pass through the hole?

Collect all results in a summary table.

ANALYSIS AND INTERPRETATION

1. Since there is no predetermined "correct" answer for this exercise, compute the percent difference of each of the six trial "muzzle velocities" from the mean velocity. Also compute the percent difference of the observed values of h_{max} from values calculated from Eq. (6.6).

2. For Case 1, Trial 1 (height of gun 4 ft above floor) calculate the velocity of the projectile just as it hits the floor. (Remember that velocity is a vector quantity.)

3. Prove mathematically that the *theoretical* range for Case 2 is a maximum when the angle of elevation is 45°.

4. The assumption in this experiment has been that only two factors affect the flight of a projectile after it has left the gun—the force of gravity and the force of air resistance. The latter was neglected in the theoretical treatment given here.

Based on collateral reading, list and discuss at least four factors besides gravity that affect the flight of *real projectiles* over long ranges in the earth's atmosphere.

NOTES, CALCULATIONS, OR SKETCHES

DATA SHEET

EXPERIMENT 6 ■ PROJECTILE MOTION

Case 1. Horizontal Projection

Trial	h	R (To MPI of three shots)	θ (By sight)	θ (Calculated)
1				
2				
3				

Case 2. Upward Projection. Baseball Player's or Surface Gunnery Problem

Trial	Angle of elevation θ	R (to MPI of three shots)	Observed y_{max} (h_{max}) trials	Observed y_{max} (h_{max}) mean
1			1. _____ 2. _____ 3. _____	
2			1. _____ 2. _____ 3. _____	
3			1. _____ 2. _____ 3. _____	

EXPERIMENT 7

FORCE AND MOTION— NEWTON'S SECOND LAW

PURPOSE

To study the relationship between force, mass, and acceleration, and to verify Newton's second law of motion in the laboratory.

APPARATUS

Precision "Air-Track" and glider with buffer stops; air blower; weight hanger and slotted weights; two Pasco "photo-gates" (Model ME-9206A or equivalent); spirit level. The glider can be loaded with varying masses, and it is pulled along the track by the force of a hanging weight. It glides with nearly zero friction on a cushion of air supplied from a blower through scores of tiny precision-drilled pinholes in the inverted V-shaped track. The glider is a true air-cushioned vehicle, or surface-effect machine (SEM). (See Fig. 7.1.)

INTRODUCTION

Force is the cause of a change in motion, or the *acceleration*, of an object that has mass. *Inertia* is that property of mass that resists a change in motion. Newton's first law of motion states that a body at rest or in motion will remain at rest or continue in motion at the same speed *and in the same direction*, unless acted on by an unbalanced force. When an unbalanced force acts on a body, the change in motion produced (i.e., the *acceleration*) is proportional to the *net* force acting and is inversely proportional to the mass of the body. This law (Newton's second law) may be stated as

$$a \propto F/m \tag{7.1}$$

where the symbol $\propto$ means "is proportional to." If the units of acceleration a, force F, and mass m are properly

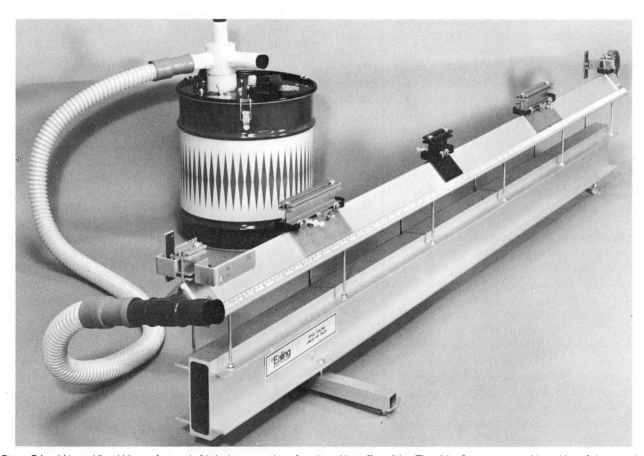

Figure 7.1. "Air track" and blower for nearly frictionless operation of an air cushion-effect glider. The glider floats on a very thin cushion of air emanating from scores of tiny, precision-drilled holes along the inverted V-shaped track. (*The Ealing Corporation*)

Table 7.1. Examples of Units Commonly Used in Newton's Second Law Calculations

Quantity	System of Units	
	SI-metric	British
Force (F)	newton (N)	pound (lb)
mass (m)	kilogram (kg)	slug (sl)
acceleration (a)	meter per s per s (m/s²)	foot per s per s (ft/s²)

chosen, the proportionality of Eq. (7.1) may be written as an equation

$$F = ma \qquad (7.2)$$

which is the mathematical statement of Newton's second law.

Table 7.1 shows two sets of units commonly used in calculations using Newton's second law.

In the SI-metric system, the *newton* (N) is the amount of force required to give a *mass* of one kilogram an *acceleration* of one meter per second per second. The units of the newton are therefore $kg \cdot m/s^2$. Similarly, in the British system, one *pound* is the amount of force required to give a mass of one *slug* an acceleration of one foot per second per second. The units of the pound (lb) are $sl \cdot ft/s^2$, and the units of the slug (sl) are $lb \cdot s^2/ft$.

An object that falls freely toward the earth (with negligible retarding force due to air resistance) experiences an acceleration of 9.81 m/s², the metric value of g, the acceleration of gravity (see Experiment 5). Equation (7.2) can be used to calculate the force required to produce that acceleration. For example, a 1-kg object falling toward the earth must experience a downward force of (1 kg × 9.81 m/s²) or 9.81 N. Similarly, a 2-kg object must experience a force of 19.62 N while falling toward the earth. The downward force experienced by a suspended or falling object, called its *weight* (w), is due to the gravitational attraction between the object and the earth. The acceleration due to gravity (9.81 m/s² or 32.17 ft/s²) is represented by the symbol g, and the *weight* of an object can be calculated using an equation similar to Eq. (7.2):

$$w = mg \qquad (7.3)$$

The weight of an object in newtons (or pounds) is equal to the product of the mass of the object in kilograms (or slugs) and the acceleration due to gravity in m/s² (or ft/s²). Of course, the object *need not be falling* to experience this force called its weight (w).

For this experiment, the test situation is basically that of Fig. 7.2. A mass $(M + m)$ is to be accelerated by a force $F = w = mg$. The force F is supplied by the weight of a hanging mass m. When the system is released it is acted upon by the unbalanced force F and it undergoes acceleration. This acceleration will be measured as accurately as possible in the experiment.

It is important to realize that the accelerating *force* results only from the weight of the hanging mass m (F = mg), but that the *mass being accelerated* is the total mass of the system, $M_T = M + m$.

For the system undergoing acceleration we can write, from Eqs (7.2) and (7.3),

$$F = w = mg = (M + m)a \qquad (7.4)$$

If the results you obtain, when substituted in Eq. (7.4) produce an identity, you have verified Newton's second law.

MEASUREMENTS

Level the track carefully and connect the air blower. Try out the glider to make sure it slides without friction. Set up the two photogates (Fig. 7.3) at convenient points along the track, exactly 1 m apart. Check the photogates for proper operation. As the glider goes through the gate, its leading edge will interrupt the light beam and start the time count; and its trailing edge will mark the reestablishment of the light beam and the end of the count. The photogate time reading is, then, the actual *elapsed time* for the glider (length L) to pass the "gate." The glider's *average velocity* during the very short time Δt_1 that it takes to pass gate 1 is given by

$$v_{1(ave)} = \frac{L}{\Delta t_1} \qquad (7.5)$$

and at the second gate,

$$v_{2(ave)} = \frac{L}{\Delta t_2} \qquad (7.6)$$

If s is the distance between gates,

$$v_2^2 - v_1^2 = 2as \qquad (7.7)$$

The acceleration of the system, a, is determined from substituting your values of v_1, v_2, and s in Eq. (7.7). Then this value of a is used with Eq. (7.4) to verify Newton's second law.

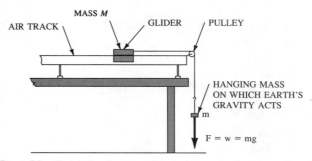

Figure 7.2. Sketch showing arrangement of apparatus for an experiment to measure uniformly accelerated motion and verify Newton's second law.

Figure 7.3. Photogates for measuring time intervals on the air track. *(Pasco Scientific Co.)*

Attach a cord or a strip of plastic tape to the glider. Its other end should hang free with a possible drop of at least a meter to the floor. Attach a weight hanger to this end. Set the photogates in position and be sure they are reset to zero. With no load on the glider, take three trials with hanging masses (m) of, say, 20, 40, and 60 g, causing accelerating forces given by $F = w = mg$.

Then place a load on the glider (say 100 g or use a longer, more massive glider) and repeat with three more trials. Be sure to record all data, including the mass of the glider and of the "weight hanger."

CALCULATIONS

For each trial determine v_1 and v_2, the average velocity of the glider as it passes gate 1 and gate 2, respectively. Then, knowing s, the distance between gates, calculate the acceleration a for each trial. Then, using the accepted value of g for your latitude and elevation above sea level, see if Eq. (7.4) results in an identity for each trial. Calculate the percent difference between the two sides of the equation, considering the left side to be the "correct" value. Show clearly that you have verified Newton's second law.

ANALYSIS AND INTERPRETATION

1. Analyze the sources of error in the performance of the experiment.

2. From collateral reading, write a paragraph on the nature of Newton's studies which led to his formulation of the laws of motion which bear his name.

3. This experiment deals with Newton's second law in terms of *force*, *mass*, and *acceleration*. However, this law is frequently referred to in terms of *impulse* and *momentum*. How can you reconcile these two approaches to the same physical law? In other words, how is $F = ma$ related to $Ft = mv$?

4. Explain the difference between uniform motion and uniformly accelerated motion.

5. If a loaded elevator has a mass of 3 metric tons, what force of tension in the hoisting cable (N) will be required to accelerated it upward at a uniform rate of 2 m/s^2?

NOTES, CALCULATIONS, OR SKETCHES

Name _____ Date _____ Section _____

Course _____ Instructor _____

DATA SHEET

EXPERIMENT 7 ■ FORCE AND MOTION—NEWTON'S SECOND LAW

Trial	Distance between photogates, s (m)	Glider length, L (m)	Glider mass, M (kg)	Added load, m (kg)	Elapsed time for glider to pass	
					Photo-gate 1, Δt_1 (s)	Photo-gate 2, Δt_2 (s)
1						
2						
3						
4						
5						
6						

COEFFICIENT OF FRICTION

PURPOSE

To study the action of frictional forces, and to determine the coefficients of kinetic (sliding) and static friction for several pairs of surfaces.

APPARATUS

Various wood and metal blocks, and surfaces on which to slide them; an inclined plane provided with a pulley at the free end; cord; weight hanger and weights.

INTRODUCTION

The surface of any material, no matter how smooth it may seem to the touch, is actually full of irregularities which oppose the sliding of any other body across it. This force of opposition as one surface slides across another is called *friction*. Friction is a force which tends to oppose the motion of two surfaces in sliding or rolling contact.

Referring to the diagram of Fig. 8.1, let A and B be two bodies, and let F be a force which tends to cause A to slide across B to the right. B is supported by a table top. Let N be the perpendicular *(normal)* component of the force experienced by block A due to its contact with block B. (In this case N equals the weight of body A.) Note that F_s, the force of static friction, acts on the bottom surface of block A and is opposite in direction to F.

If F is zero, F_s is also zero. As F is increased, F_s increases also, until the condition is reached where motion impends. At this instant, just as the block A would begin to slide, the force of static friction reaches its maximum or

limiting value. This value is said to be the *force of limiting friction* or the *maximum force of static friction $F_{s_{max}}$*.

Once the body A begins to move, it will be found that the force of friction diminishes somewhat. This lowered value of frictional force for surfaces where sliding already exists is called the *force of kinetic (sliding) friction F_k*. For steady motion without acceleration, $F = F_k$, for the horizontal surfaces shown.

For elementary studies of friction in the laboratory, the following statements are nearly, if not quite exactly, true for dry surfaces:

1. The force of sliding friction is almost independent of the area of contact, but is directly proportional to the *normal force*.

2. The force of kinetic (sliding) friction depends on the nature of the two surfaces.

3. The force of kinetic (sliding) friction is almost independent of the relative velocity of the sliding surfaces for normal velocity ranges.

Coefficients of Friction

In terms of the concepts defined above, the *coefficient of kinetic friction* is defined as

$$\mu_k = \frac{\text{force of kinetic friction}}{\text{normal force}}$$

or

$$\mu_k = \frac{F_k}{N} \tag{8.1}$$

For *static,* or *limiting* friction, the coefficient is

$$\mu_s = \frac{F_{s_{max}}}{N} \tag{8.2}$$

Coefficients of friction have no units, since they are ratios of two forces.

MEASUREMENTS

1. Weigh the several blocks, and record the nature of their materials, their surfaces, and their weights. *Note:* For convenience it is suggested that metric units be used—all weights and forces in newtons.

2. Put the inclined plane (Fig. 8.2) with its base in a *horizontal* position with the pulley projecting beyond the table edge. Place a wooden block on the plane with its largest face in contact with the plane. (See Fig. 8.2.) Connect a

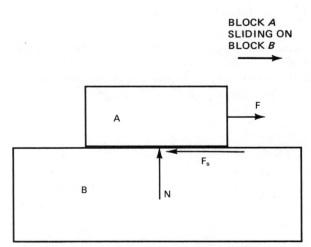

Figure 8.1. Diagram showing the factors involved in static friction—F less than $F_{s_{max}}$.

Figure 8.2. Impending motion down an inclined plane with friction. When the block just starts to slide, $\mu_s = \tan \theta$. *(Central Scientific Co.)*

cord from the block, over the pulley to a weight hanger. Set a mass on the block, say 500 g, to increase the value of N.

3. Determine the force required to just start the block sliding (force of maximum static friction, $F_{s_{max}}$). Then determine the force required to maintain a *steady, slow velocity* (force of kinetic friction, F_k.) (Start the block with a gentle push.)

4. Now turn the block on a side with smaller area and repeat all of Step 3.

5. With the same block, on the side with largest area, but with the cord and 500-g mass removed, slowly elevate the plane until the block just *starts* to slide of its own accord. At the *instant* of starting, the forces diagrammed in Fig. 8.3a are in equilibrium, where $F_{s_{max}}$ is the maximum force of *static* friction. Consequently these forces, if drawn as vectors (Fig. 8.3b), would form a closed triangle, as

$wF_{s_{max}}N$. But triangle $wF_{s_{max}}N$ is similar to the triangle formed by the plane itself; therefore,

$$\mu_s = \frac{F_{s_{max}}}{N} = \tan \theta = \frac{h}{b} \qquad (8.3)$$

Read and record the angle of elevation, θ, of the plane. This angle at the very point of impending motion is called the *angle of slip*.

6. Repeat Steps 2 through 5 with at least three other sets of sliding surfaces, as metal on wood, steel on steel, rubber on concrete, etc.

7. Record all data on the Data Sheet.

CALCULATIONS

1. Remember that all *masses* used must be converted to *forces* (weights) by the relation

$$w = mg \qquad (8.4)$$

Force (newtons) = mass (kg) $\times$ 9.81 m/s^2.

2. For each pair of surfaces compute the coefficient of kinetic friction μ_k from Eq. (8.1). Note differences, if any, caused by a change in the area of contact between the surfaces.

Calculate the percent error of your results from the accepted values found in tables.

3. Calculate the coefficient of static friction μ_s for each pair of surfaces from Eq. (8.3). Compare the results with the tangent of angle θ.

Summarize all results in a neat table so that your answers may be readily compared with accepted values for the surfaces used.

ANALYSIS AND INTERPRETATION

From collateral reading on the subject of friction, discuss the following topics:

1. The effect of very high relative velocity of the sliding surfaces on the force of sliding friction.

2. The theory of lubrication as it is related to friction.

3. The concept of rolling friction, as contrasted with sliding (kinetic) friction.

Answer the following questions:

1. Why is it necessary that the block be moving with uniform velocity, in determining the force of kinetic friction?

2. Will a speeding automobile stop more quickly with the wheels locked (tires skidding) or with the wheels braked just to the point of impending skidding? Why?

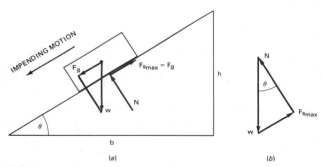

Figure 8.3. *(a)* Space diagram and *(b)* vector diagram of the forces acting on the block at the moment of impending motion down an inclined plane.

Name _____ Date _____ Section _____

Course _____ Instructor _____

DATA SHEET

EXPERIMENT 8 ■ COEFFICIENT OF FRICTION

Description of sliding surfaces (such as rubber on hardwood)		Total weight of sliding object (N)	Force of static friction $F_{s_{max}}$ (N)	Force of kinetic friction F_k (N)	Angle of plane for impending motion (degrees)
	Large area				
	Small area				
	Large area				
	Small area				
	Large area				
	Small area				
	Large area				
	Small area				

EXPERIMENT 9

SIMPLE MACHINES BASED ON THE LEVER

PURPOSE

To study the operation of some simple machines, and to evaluate their actual mechanical advantages and efficiencies.

APPARATUS

Assorted pulleys and cord; commercial block and tackle (small size); wheel and axle; screw jack; differential pulley (chain hoist); gear train; weight hanger and weights; spring balance; meter stick.

INTRODUCTION

The word *machine* conveys many meanings, depending on the context in which it is used. In physics and technology, however, a rather restricted meaning is adopted. A *machine* is defined as a *device for the advantageous application of a force*.

Most machines of industry are very complicated devices, but no matter how complex they become, they are just combinations of only three basic machines—the *lever,* the *inclined plane,* and the *hydraulic press.*

Among the simple machines which operate on the lever principle are *pulleys, gears,* the *wheel and axle,* the *block and tackle,* and the *differential pulley.*

The *screw jack,* the *wedge,* and the *cam* are examples of the inclined plane. The inclined plane will be used in Experiment 10 for a study of the *work-energy theorem.*

In this experiment we shall study the following simple machines: pulley systems, including block and tackle; the differential pulley; the wheel and axle; the gear train; and the screw jack.

Machines may be used to multiply the force applied (input force, F_i), thus delivering a force output F_o, greater than the force input; or they may be so designed as to multiply the distance or speed, so that the load moves farther (or faster) than does the effort force applied. The input distance is designated by S_i; the output distance by S_o.

It must be emphasized, however, that a machine which multiplies force will sacrifice speed and distance, and that one which multiplies speed (or distance) will require a force input larger than the force output. In any case, the law of conservation of energy must be obeyed. For a machine this means that the *work output* cannot be greater than the *work input.*

In the theoretical (ideal) machine with no losses, work output would be equal to work input:

$$F_i S_i = F_o S_o \tag{9.1}$$

where, as above,

F_o = output force (load)
S_o = distance output force acts
F_i = input force (effort)
S_i = distance input force acts

Such an ideal is never achieved, however, and the extent to which the ideal is approached is called the *efficiency* of the machine.

$$\text{Efficiency} = \frac{\text{work output}}{\text{work input}} = \frac{F_o S_o}{F_i S_i} \tag{9.2}$$

The *actual* mechanical advantage of a machine is

$$\text{AMA} = \frac{\text{force output}}{\text{force input}} = \frac{F_o}{F_i} \tag{9.3}$$

From Eqs. (9.1) and (9.3) it can be readily deduced that for the ideal machine the *theoretical* mechanical advantage is

$$\text{TMA} = \frac{\text{input distance}}{\text{output distance}} = \frac{S_i}{S_o} \tag{9.4}$$

From Eqs. (9.2) to (9.4) it is seen that

$$\text{Efficiency} = \frac{\text{actual mechanical advantage}}{\text{theoretical mechanical advantage}} \tag{9.5}$$

$$\text{or, Eff.} = \frac{\text{AMA}}{\text{TMA}}$$

TMA varies from one machine to another, being expressed in terms of the geometry of the machine. The AMA of *all machines* is given by the same expression, i.e., the ratio of output force to input force.

MEASUREMENTS

(Metric system only. Remember that *mass* loads are converted to *weights* and *forces* by the relation $w = F = mg$).

Part I. Pulley Systems

(See Fig. 9.1 for illustrations of pulley systems.)

1. Rig a single *fixed* pulley on a stand and run the cord through it. Use a 500-g (mass) load, and determine the input force required to raise the load *at (constant) uniform speed.* How far does the load move when the input force moves 30 cm?

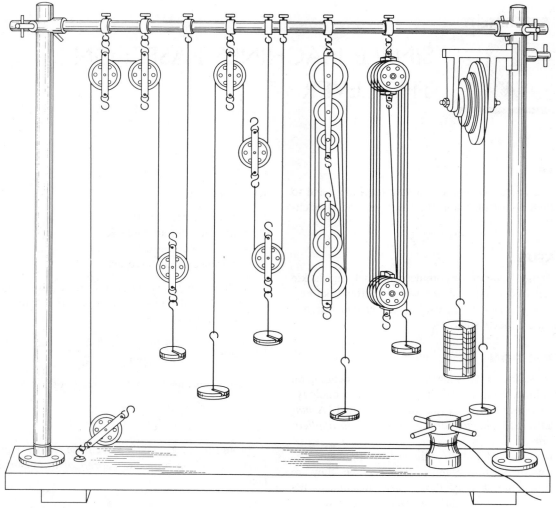

Figure 9.1. A laboratory display for the study of pulleys and pulley systems. At the far right are a wheel-and-axle and a capstan. *(Sargent-Welch Scientific Co.)*

2. Rig a single *movable pulley,* and load it with a 500-g mass. Determine the input force required to raise the load at a uniform speed. How far does the load move when the input force moves 30 cm?

3. Rig a *multiple* pulley system with three sheaves (pulleys) on each block. Use a load of 2 kg. Determine the input force and the distance the load moves for a 30-cm input force distance, for two cases:

a. End of cord tied to fixed block

b. End of cord tied to movable block

Make a note of the number of strands of cord attached to the movable block in each case.

4. Rig a *commercial block and tackle,* if available, using a load in excess of 30 kg. Determine the input force and the distance the load moves for a 1-m movement of the input force. Note the number of ropes to the movable block.

Part 2. Differential Pulley (Chain Hoist)

(See Fig. 9.2.)

1. Determine the radii of both the smaller and the larger pulleys.

Figure 9.2. Differential pulley or "chain fall." *(Sargent-Welch Scientific Co.)*

2. Determine the input force required to raise a load of 20 kg. How far does the load move when the input force moves 5 m?

Part 3. Wheel and Axle

(See far right Fig. 9.1.)

1. Clamp the apparatus to a heavy laboratory stand. Wind a cord several times around the axle, and another cord several times around the wheel *in the reverse direction*. Apply a load of 5 kg to the cord on the axle, and determine the force required (applied to the wheel cord) to lift the load at uniform speed.
2. Measure the radii of both wheel and axle.
3. How far does the load move for each 1-m movement of the input force?

Part 4. Gear Train

(See Fig. 9.3.)

1. Clamp the apparatus to a heavy stand. Attach a load of at least 2 kg to one end of the gear train, and a cord with weight hanger at the other. (The arrangement should multiply the force.) Determine the force required to lift the load at a uniform speed.
2. Record the number of gear wheels and the number of teeth on each.
3. How far does the load move for a 1-m motion of the applied force?

Part 5. Screw Jack

(Fig. 9.4 shows a small model.)

1. Put a heavy load on the jack. (From 10 to 100 kg, depending on the size of the jack available.) Note care-

Figure 9.4. Laboratory model of a screw jack. *(Central Scientific Co.)*

fully the force required at the end of the jack handle to lift the load at uniform speed. Be sure the force is applied perpendicular to the jack handle.
2. Measure very accurately the pitch of the screw and the length of the jack handle from the center of rotation to the point where the force is applied.

CALCULATIONS AND RESULTS

For each of the simple machines tested, calculate the AMA, the TMA, and the efficiency.

It will be necessary for you first to derive (or look up in your physics textbook) the expression for the TMA of the wheel and axle, the differential pulley, the gear train, and the screw jack, expressed in terms of the geometry of the machine.

Display all your results in a neat table.

ANALYSIS AND INTERPRETATION

1. Explain carefully what you have learned about machines from this experiment with respect to what they can do for forces, for distance and speed, and for work.
2. Why is AMA the same expression for all machines, whereas the expression for TMA is different for each machine?
3. State the principle of work as it applies to machines.
4. Suppose you had a block and tackle consisting of two blocks with two sheaves in each block. How could you rig the system for the greatest possible force to act on the load (movable block)? What would be the TMA? Make a neat sketch of the setup you would use.
5. Why are screw jacks intentionally designed for a low efficiency?

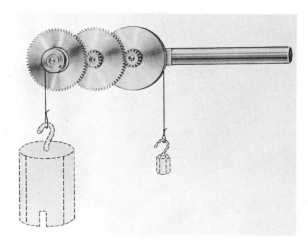

Figure 9.3. Gear train made up of spur gears. *(Sargent-Welch Scientific Co.)*

NOTES, CALCULATIONS, OR SKETCHES

DATA SHEET

EXPERIMENT 9 ■ SIMPLE MACHINES BASED ON THE LEVER

Note to student: Devise your own neat tables for recording data, under the headings below.

Part 1.   Pulley Systems

Part 2.   Differential Pulley

Part 3.   Wheel and Axle

Part 4.   Gear Train

   Sketch of gear train with number of teeth on each gear indicated.

Part 5.   Screw Jack

NOTES, CALCULATIONS, OR SKETCHES

THE INCLINED PLANE AND THE PRINCIPLE OF WORK

PURPOSE

To study the principle of work and the effect of friction, using the inclined plane as a simple machine.

APPARATUS

Inclined plane fitted with a fixed pulley and a protractor for measuring the angle of the incline; Hall's car; friction blocks of various materials; meter stick; hanger and set of masses; strong cord.

INTRODUCTION

A machine is a device for the application of a force in such a manner as to accomplish useful work. Machines may multiply force at the expense of distance (or speed), or they may multiply speed at the expense of force, but the product of force times distance (work output) cannot be greater than the product of force times distance (work input). This generalization is a form of the law of conservation of energy, and applied to machines, it is known as the *principle of work*.

This experiment deals with just one of the basic machines—the inclined plane. Its mechanical advantage and its efficiency will be determined and frictional forces will be studied. The investigations should also verify the principle of work. The use of vectors in the analysis of forces is also illustrated.

Remember that *forces* must be expressed in newtons (N). If you use *masses* for loads (kg and g), these values must be converted to newtons, recalling that $w = F = mg$.

Figure 10.2. Inclined plane and Hall's carriage setup for study of the law of conservation of energy (principle of work). Friction assumed negligible. *(Central Scientific Co.)*

$$\text{Force (N)} = \text{kg} \times 9.81$$
$$\text{g} \times 0.00981$$

Analysis of Fig. 10.1 shows that if a rolling car of mass m is resting on a plane inclined at an angle θ with the horizontal and *if friction is neglected*, the force which tends to start it rolling down the plane is

$$F = mg \sin \theta \text{ (in mks units)} \qquad (10.1)$$

and a force $F' = F$ would be required to hold it in place (or to pull it up the plane at a uniform velocity, if friction is absent).

Now $\sin \theta = h/L$ and, therefore, from Eq. 10.1,

$$\frac{F}{w} = \sin \theta = \frac{h}{L} \quad \text{or} \quad F = \frac{w \times h}{L} = \frac{mgh}{L} \qquad (10.2)$$

The *actual mechanical advantage* (AMA) is

$$\text{AMA} = \frac{w}{F} = \frac{mg}{F} \qquad (10.3)$$

and the *theoretical mechanical advantage* (TMA) is

$$\text{TMA} = \frac{L}{h} \qquad (10.4)$$

Now the work done in pulling the car up the plane (see Fig. 10.2) is

$$\text{Work} = \text{force} \times \text{distance}$$
$$W = F' \times L \qquad (10.5)$$

Note that W denoted *work; weight* is denoted by w. The

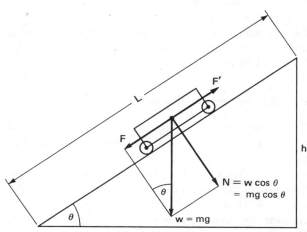

Figure 10.1. Analysis of forces on an inclined plane—friction assumed negligible.

principle of work requires that work done on the car to pull it up the plane (friction assumed absent) should be equal to its gain in gravitational potential energy, or

$$GPE = mgh \qquad (10.6)$$

This gain in GPE is expressed in newton-meters (N · m), or joules (J).

Although friction has been assumed negligible in the foregoing, there is, of course, some friction present even in well-lubricated wheels, and the equation $F'L = w \times h$ will not reduce to a perfect identity when data from the experiment are substituted in it. The *efficiency* of the inclined plane and Hall's car combination may be found from the relation

$$\text{Efficiency} = \frac{\text{work output}}{\text{work input}}$$

or

$$\text{Efficiency} = \frac{wh}{F'L} = \frac{mgh}{F'L} \qquad (10.7)$$

Figure 10.3 illustrates the situation in which *friction is not neglected.* Let w be the weight of the object resting on the inclined plane. It will *tend to slide* down the plane under the action of the force F_g, which is the component in the line of the plane of the force of gravity on the block. From the vector diagram, $F_g = w \sin \theta = mg \sin \theta$ (in mks units), as before.

If, now, a force F' is applied to pull the object up the plane, the value of F' must be that required to overcome not only the F_g component of the force of gravity but also the force of kinetic friction, F_k, between the block and the surface of the plane. Expressed as an equation,

$$F' = F_g + F_k \qquad (10.8)$$

but

$$F_g = w \sin \theta$$

and

$$F_k = \mu_k N = \mu_k w \cos \theta$$

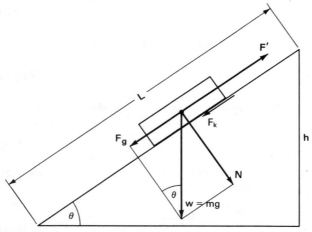

Figure 10.3. Analysis of forces on an inclined plane when friction is taken into account.

where N is the component of the force of gravity *normal* to the plane. Consequently,

$$F' = w \sin \theta + \mu_k w \cos \theta \qquad (10.9)$$

or

$$F' = w(\sin \theta + \mu_k \cos \theta)$$
$$= mg(\sin \theta + \mu_k \cos \theta) \qquad (10.10)$$

The law of conservation of energy states that, in any machine, the useful work output is equal to the energy input minus any losses within the machine (such as those from doing work against friction). For machines, this law is known as the *principle of work.*

For the case being studied, *the principle of work* may be expressed by

$$wh = F'L - F_kL$$

or

$$wh = F'L - \mu_k w \cos \theta \, L$$

Factoring,

$$wh = L(F' - \mu_k w \cos \theta)$$

or

$$mgh = L(F' - \mu_k mg \cos \theta) \qquad (10.11)$$

MEASUREMENTS

Note: Use the metric system—forces in newtons (N), masses in kg or g; distances in m or cm.

Part 1. Friction Neglected

Determine the mass of the Hall's car and record. Adjust and oil the wheels. Set the plane at a 30° angle, load the car with a 500-g mass, and determine F', the force required to pull the loaded car up the plane at uniform velocity (without acceleration) (see Fig. 10.2). Record the average of several values of F', for the car loaded with 500 g. Take two more trials, with a different load in the car each time. Take measurements of h, L, and θ.

Repeat with the plane set at a 45° angle. Take care that the cord which pulls the car is parallel to the plane and that the pulley over which the cord passes is well oiled. Try to keep friction at a minimum.

Part 2. Inclined Plane with Friction

Set the plane at a 20° angle. Determine the masses of the test blocks and record. Note the nature of the sliding surfaces (i.e., wood on wood, rubber on wood, etc.) Put one of the test blocks on the plane, and load the "weight" hanger until the block just slides up the plane without acceleration. (*Hint:* Start it sliding with a gentle push.) Record the values of F', $w (= mg)$, and θ, and also h and L. For another trial, set angle θ at 40°.

Repeat with blocks of different materials.

CALCULATIONS

Part 1. Friction Neglected

1. Calculate F for all trials from Eq. (10.2). Note that the experimental value F' differs slightly from the theoretical value F.

2. Calculate from Eq. (10.5) the work done on the car as it was pulled up the plane. Compare this with the theoretical gain in GPE calculated from Eq. (10.6). Compute the efficiency of the inclined plane and Hall's car from Eq. (10.7).

Part 2. Inclined Plane with Friction

1. From your measurements and Eq. (10.9), calculate the coefficient of kinetic friction μ_k for each of the sliding pairs of surfaces. Compare your results with accepted values from tables.

2. From Eqs. (10.10), (10.5), (10.6), and (10.7), using values of F', L, h, w, and θ as measured by experiment, calculate the overall efficiency of the inclined plane for each of the blocks used.

3. Using the values of μ_k from Part 2(1) above, verify the *principle of work* as expressed by Eq. (10.11).

ANALYSIS AND INTERPRETATION

In addition to the usual discussion of errors and practical applications, be sure to do the following:

1. State the general law for the TMA of an inclined plane. Explain the difference between TMA and AMA. Why is AMA expressed in the same way for all machines, whereas TMA has a different expression for each machine?

2. If a 1500-kg automobile is being driven at 20 km/h up a grade which rises 15 m for every 30 m along the incline, calculate the minimum power its engine is delivering to the driving wheels, in kW. Neglect friction.

3. List and discuss at least three machines or parts of machines which are based on the inclined plane principle.

4. Discuss ways in which friction must be taken into account in the design of machines.

5. Mention several examples of inclined planes being used as a component of machine design. For each example explain the factors that influence the selection of an appropriate angle for the incline.

NOTES, CALCULATIONS, OR SKETCHES

DATA SHEET

EXPERIMENT 10 ■ THE INCLINED PLANE AND THE PRINCIPLE OF WORK

Part 1. Friction Neglected

Angle of plane (θ)	Trial	Mass of car (kg)	Load in car (kg)	F' (N)	h (m)	L (m)
30°	1					
	2					
	3					
45°	1					
	2					
	3					

Part 2. Inclined Plane with Friction

Trial	Nature of sliding surfaces (e.g., wood on wood)	Mass of block (kg)	Angle of plane (θ)	F' (N)	h (m)	L (m)
1			20°			
			40°			
2			20°			
			40°			
3			20°			
			40°			

NOTES, CALCULATIONS, OR SKETCHES

EXPERIMENT 11

UNIFORM CIRCULAR MOTION—
CENTRIPETAL FORCE

PURPOSE

To study centripetal force as the result of uniform circular motion, and to verify Newton's laws with respect to circular motion.

APPARATUS

Centripetal force apparatus; variable speed rotator with revolution counter; precision stop clock or stopwatch, vernier caliper, laboratory stand and metric masses; spirit level.

INTRODUCTION

Newton's second law of motion applies to curvilinear motion as well as to straight-line motion. Circular motion is one form of curvilinear motion. Force, mass, and acceleration are related in uniform circular motion, and one purpose of this experiment is to determine how they are related. If a body is moving with constant speed in a circular path, there must be a force directed toward the center of the circle which acts to continually change the *direction* of motion of the body and keep it in its circular path. This force is called *centripetal* ("center-seeking") *force*.

A body that moves with constant speed in a circular path has a velocity whose *magnitude* is constant, but whose *direction* is continually changing. Since velocity is a vector quantity, a change in direction means a change in the *velocity*, even though the magnitude, or *speed*, is constant. Time rate of change of velocity is *acceleration*, and in this experiment the direction and magnitude of this acceleration will be studied.

Figure 11.1a is a *space* diagram. Let B represent a body of mass m which is moving in a circular path of radius **r**. Its speed is constant, but its direction of motion is continually changing. The vector **v** represents the instantaneous magnitude and direction of the velocity of the body at point B.

A short time (Δt seconds) later, the body is at B', and its velocity is represented by the vector **v'**, whose *magnitude* is equal to that of vector v, but whose *direction* is different. In the time interval Δt the body has moved along the circle a path distance $s = v\Delta t = r\theta$ (θ in radians).

Figure 11.1b is a *vector* diagram of velocities, with **v** and **v'** plotted from a common point A, parallel to their directions as given in Fig. 11.1a. The vector $\Delta v = bb'$, which closes the vector triangle, is the velocity vector which, when added vectorially to **v**, gives **v'**. As a vector equation $v + \Delta v = v'$. The vector Δv represents a change

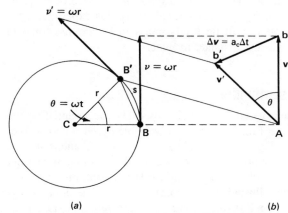

Figure 11.1. (a) Space diagram showing the factors involved in uniform circular motion. (b) Vector diagram showing the difference between two velocity vectors. Vector Δv represents the *directional* change in velocity from v to v'. But change in velocity is equal to acceleration multiplied by time; therefore, vector Δv equals $a_c\Delta t$ in *magnitude*.

in velocity, and it is equal to the product of the *centripetal acceleration a_c* times the time interval Δt. (The *magnitude* of $\Delta v = a_c\Delta t$.) Now if the time interval Δt is very short, the arc distance s (Fig. 11.1a) is essentially equal to the chord BB', and triangle CBB' in Fig. 11.1a is similar to triangle Abb' of Fig. 11.1b. Therefore, *in terms of magnitudes,*

$$\frac{a_c\Delta t}{v} = \frac{v\Delta t}{r}$$

or simplifying,

$$a_c = \frac{v^2}{r} \tag{11.1}$$

Since $v = \omega r$ (where ω = angular velocity in radians per second),

$$a_c = r\omega^2 \tag{11.2}$$

Newton's second law for *translation* is generally expressed as follows:

$$F = ma \quad (F \text{ in newtons})$$

For uniform circular motion the second law of motion is

$$F = ma_c$$

Substituting from Eq. (11.1),

$$F = \frac{mv^2}{r} \qquad (11.3)$$

and, substituting from Eq. (11.2),

$$F = mr\omega^2 \qquad (11.4)$$

Equation (11.4) is the basic equation for *centripetal force,* the force required to hold a mass *m* in a circular path of radius *r* when it is revolving around the center with constant angular velocity ω. Just as with Newton's laws for straight-line motion, *absolute units* of force must be used with Eqs. (11.3) and (11.4). In this experiment, SI-metric units will be used—mass in kilograms, force in newtons, and distance in meters.

MEASUREMENTS

The centripetal force apparatus is shown mounted on the rotator in Fig. 11.2, and the variable speed rotator is pictured in Fig. 11.3. The mass *m* is cylindrical and is cradled by three support guides *G,* so that it may move in and out along the axis of the spring *Z*. The tension on the spring may be adjusted by the knurled nut *K*. The apparatus is rotated in a horizontal plane around a vertical axis, and as it rotates at higher angular speeds the stretched spring provides the centripetal force to hold mass *m* in a circular orbit. A pointer *P* is so designed that when the cylindrical mass presses outward against it, it pivots and moves the pointer across the index *I*. The index is on the axis of rotation and the pointer's position can be clearly seen (due to the persistence of vision of the human eye) while the apparatus is rotating.

During a trial run the speed of the rotator is varied by

Figure 11.3. Variable speed rotator. *(Central Scientific Co.)*

means of a friction drive plate until the mass *m* moves out far enough to make the pointer *P* steady down opposite the index *I*. The number of revolutions for a stipulated time interval are then determined.

Adjust the spring tension to near minimum (scale *S*). Mount the centripetal force apparatus on the rotator, and level the apparatus carefully with the spirit level. Start rotating at low speed, and with the eyes on a level with the index, slowly increase rotational speed until the rotating cylindrical mass moves out, touches the pointer, and brings the pointer opposite the index. Some practice will be required to develop skill in keeping the pointer on the index. Record the reading on the revolution counter, which up to now, has not been engaged. Then engage the counter and start the time clock simultaneously, all the while making the necessary minor adjustments of the speed control to keep the pointer on the index. At the end of (exactly) 1 min, disengage the counter and record the reading. Do not allow the counter to coast after its gear has been disengaged. Conduct ten 1-min trials in succession, recording the readings of the revolution counter at each interval.

Remove the apparatus from the rotator, and suspend it on a heavy laboratory stand, as shown in Fig. 11.4. Determine the force necessary to cause the pointer *P* to point directly at the index *I*. Be sure to include the weight of the cylindrical mass itself and that of the weight hanger, in obtaining the total force required to stretch the spring the same amount that it was stretched while the mass was revolving. Express this force in newtons. (Remember, $w = mg$.)

Measure the radius of rotation of the cylindrical mass with a vernier caliper *while it is still in the equilibrium position* on the laboratory support. This is the distance from the vertical axis *OO* to the *center line* of the cylindrical mass (express in meters).

Now, adjust the spring tension to near maximum by turning the knurled nut *K*, and repeat the entire procedure.

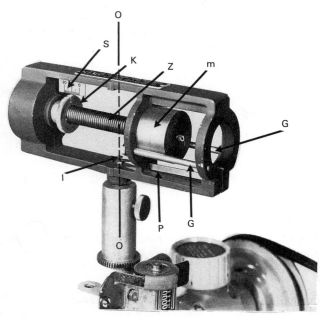

Figure 11.2. Centripetal force apparatus shown mounted on a variable-speed rotator. The important components of the apparatus are labelled for identification. *(Central Scientific Co.)*

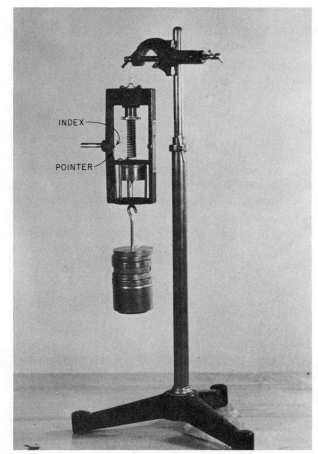

INDEX

POINTER

Figure 11.4. Centripetal force apparatus vertically mounted, to determine the gravitational force equivalent of the centripetal force.

CALCULATIONS

For both sets of data (minimum spring tension and maximum spring tension), calculate an average value of the angular velocity ω in radians per second, from the relation $\omega = (2\pi N)/60$, where N is revolutions per minute.

From Eq. (11.4) compute the value of the centripetal force for each of the two trials. Compare these values with the forces required to produce the same displacement of the spring for the two trials. Compute the percent difference for each trial. Display your results in such a way that they can be readily interpreted. Show how Newton's second law for uniform circular motion is verified.

ANALYSIS AND INTERPRETATION

1. Discuss significant errors, if any. What factors cause the errors?

2. What have you proved by your results? Be sure to explain how you have dealt with the purpose of the experiment.

3. Explain carefully, in the light of Newton's laws, how centripetal force holds revolving objects in orbit.

4. Explain the seeming contradiction that in Eq. (11.3) centripetal force is inversely proportional to the radius, while in Eq. (11.4) it is directly proportional to the radius. (Use dimensional analysis.)

5. A satellite is to orbit the earth in a circular path at an altitude of 1000 mi. What must be its linear speed in miles per hour? Earth's radius is 4000 miles. (*Hint:* What is the value of g at that distance from the earth's surface?)

6. An experiment in "weightlessness" is being conducted, using a conventional airplane. If the plane is executing an outside loop at a speed of 700 km/h, what should the radius of the loop be (in meters) in order that the passengers would just "float" within the cabin of the plane? (Consult your physics textbook for problems 5 and 6.)

DATA SHEET

EXPERIMENT 11 ■ UNIFORM CIRCULAR MOTION—CENTRIPETAL FORCE

Elapsed time (min)	Trial 1 (minimum tension)	Trial 2 (maximum tension)
	Revolution counter reading	Revolution counter reading
0		
1		
2		
3		
4		
5		
6		
7		
8		
9		
10		
Force (weight) required to stretch spring to same position $(w = mg)$	_____ N	_____ N

Mass of revolving cylinder _____ kg

Radius of the circular motion _____ m

SIMPLE HARMONIC MOTION— THE PENDULUM AND THE SPIRAL SPRING

PURPOSE

To study simple harmonic motion by means of (1) a pendulum and (2) a spiral spring; and to use a pendulum for the laboratory determination of g, the acceleration of earth's gravity.

APPARATUS

Metal sphere and wooden sphere, both of the same diameter; strong thread; long support rod and clamp; 2-m stick; stop clock; spiral spring; weight holder and slotted weights (or a Jolly balance).

INTRODUCTION

Any motion that repeats itself in equal intervals of time is called *periodic motion*. The simplest kind of periodic motion is perhaps *uniform circular motion*. Other common examples of simple periodic motion are: the swinging pendulum of a clock, the oscillation of a loaded spiral spring, and the vibration of a tuning fork.

A special form of simple periodic motion called simple harmonic motion will be studied in this experiment. *Simple harmonic motion* (SHM) is defined as cyclical or oscillatory motion in which the resultant force on the oscillating body at any instant is directly proportional to its displacement from the "rest position," and opposite in direction to the motion of the body at that instant. As an equation, SHM is defined by

$$F = -kx \tag{12.1}$$

where F is the resultant force on the oscillating body, x is the displacement, and k is a constant. Now, when an unbalanced force is exerted on a body, the body undergoes acceleration. Since, from Newton's second law, $F = ma$, from Eq. (12.1), $ma = -kx$, or the acceleration of a body of mass m experiencing SHM is

$$a = -\frac{k}{m}x \tag{12.2}$$

As a special case we consider SHM as the projection on a diameter, of a point which is moving in a circular path at uniform speed. Figure 12.1 shows the relationships involved. The point P is moving counterclockwise around the circle of radius r with a speed v. If for every position of P a perpendicular is drawn to the diameter AB, a series of points P' will be determined. As P moves around the circle, P' moves back and forth across the diameter AB. P' is experiencing *simple harmonic motion*, while P experi-

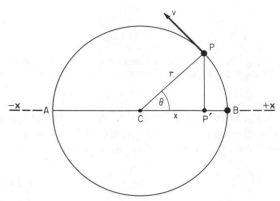

Figure 12.1. The projection on a diameter of a particle in uniform circular motion. The particle P is in uniform circular motion around the point C. Point P' is at all times the projection of particle P on the diameter AB. The point P' describes simple harmonic motion (SHM).

ences uniform circular motion. The *displacement* at any instant during a cycle is defined as the distance x from the center C to the point P'. The displacement x varies from a value of zero at C to a maximum value of r.

The point P', as it oscillates along the diameter between B and A, has its greatest *velocity* as it moves through C. It experiences negative acceleration on the way from C to A and momentarily has zero velocity at A, as it reverses course for the return trip through C to B. Its *maximum velocity*, as pointed out above, occurs at C; and its *maximum acceleration* occurs at A and B, the points where it reverses direction. Note that Eq. (12.2) predicts this result, since r is the largest value x can have.

The period T of the simple harmonic motion is the time required for one *complete* oscillation. *Frequency* is the number of complete oscillations per second.

Derivation of Formula for Period T Referring to Fig. 12.2, it is seen that the point P, whose speed is v, will travel a distance $2\pi r$ in a time T where T is the time (s) required for a complete revolution. Therefore

$$T = \frac{2\pi r}{v} \tag{12.3}$$

The centripetal acceleration of P toward the center is $a_c = v^2/r$. Its component along AB is $a = v^2/r \times \cos\theta$. But $\cos\theta = x/r$, and so the *magnitude* of a at any instant is given by

$$a = \frac{v^2}{r} \cdot \frac{x}{r} = \frac{v^2}{r^2}x \tag{12.4}$$

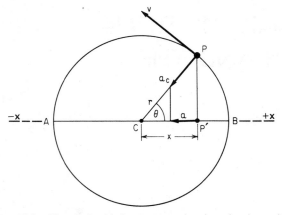

Figure 12.2. The relationship between centripetal acceleration and the linear acceleration of a particle in simple harmonic motion.

However, since the acceleration vector **a** points in the negative x-direction when x is positive and in the positive x-direction when x is negative, as a *vector,*

$$\mathbf{a} = -\frac{v^2}{r^2}x \tag{12.5}$$

Eq. (12.5) demonstrates that such a motion conforms to the definition of SHM as given by Eq. (12.2). Note that the quantity v^2/r^2 is actually the constant k/m of Eq. (12.2). Transposing (Eq. 12.4), and taking the square root,

$$\frac{r}{v} = \sqrt{\frac{x}{a}}$$

Substituting in Eq. (12.3) a formula for the period of the simple harmonic motion (SHM) is obtained.

$$T = 2\pi \sqrt{\frac{x}{a}} \tag{12.6}$$

Part I. The Pendulum

To apply this analysis to the special case of the simple pendulum, refer to Fig. 12.3. The downward pull of gravity, $w = mg$, on the pendulum ball has been resolved into two components, one (F_P) *parallel* to the cord L, the other (F_T) perpendicular to it (*tangent* to the circular arc of the pendulum). (Throughout this derivation, and inherent in the resulting equation for the period of a simple pendulum, is the assumption that x is equal to the arc distance Cm. This condition is essentially true only for very small angles up to about 4°.) From the similar triangles of Fig. 12.3,

$$\frac{F_T}{w} = \frac{x}{L} \tag{12.7}$$

Since the component of w parallel to the cord L (F_P) is balanced by an oppositely directed, equally strong tension force in the cord, the component of w perpendicular to the cord (F_T) is the resultant force acting on the pendulum bob. According to Newton's second law ($F = ma$), the resultant force acting on the bob must be equal to the mass of the bob times its acceleration. That is, for the bob,

$F_T = ma$. Substituting this expression and $w = mg$ in Eq. (12.7) gives $(ma)/(mg) = x/L$, from which $x/a = L/g$. Substituting this result in Eq. (12.6) gives

$$T = 2\pi \sqrt{\frac{L}{g}} \tag{12.8}$$

an expression for the period of a pendulum in terms of its length—a measurable quantity. Note that the period is independent of the mass of the pendulum bob. Keep in mind that the formula will give acceptable results only for small amplitudes ($\theta = 4°$ or less). T will be in seconds when L is in meters and g is in m/s².

MEASUREMENTS

Set up a very solid laboratory support and pendulum clamp. (See Fig. 12.4.) Attach a wooden sphere to the thread, and clamp the thread to the support so that the length of the pendulum (*to center of sphere*) is *exactly* 2 m. Displace the sphere to one side through an angle of not more than 4°, and release, allowing the pendulum to oscillate.

1. Record the time required for 50 *complete* oscillations.
2. Repeat with a metal sphere and the same pendulum length.

With the metal sphere, repeat the procedure with successive pendulum lengths of 150, 100, and 50 cm (Trials 3, 4, and 5). (Pendulum length measurements should be accurate to 0.5 mm.)

Finally, with the metal sphere and a length of 1 m, determine the time required for 50 complete oscillations when the pendulum bob is displaced about 10, 20, and 40° from the rest position. (Trials 6, 7, and 8.)

Record all data on the Data Sheet.

CALCULATIONS AND RESULTS

1. Compute the period T for each trial from your experimental data, to three significant figures.

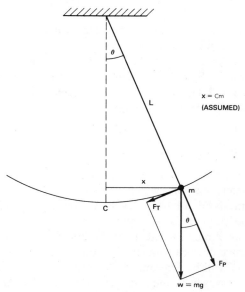

Figure 12.3. Analysis of the forces acting on a simple pendulum.

Figure 12.4. Pendulum and laboratory timer.

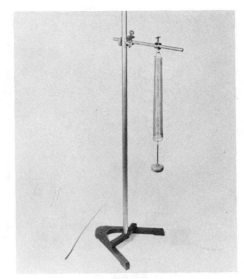

Figure 12.5. Spiral spring loaded for observations of simple harmonic motion.

2. By comparing the results of Trials 1 and 2, can you verify the finding implicit in Eq. (12.8)—that the period of a pendulum is independent of the mass of the pendulum bob?

3. Calculate the square of the period for each observation, Trials 2 to 5, inclusive.

4. Calculate the value of g from each measurement, Trials 2 through 5, inclusive. Compute the mean, and compare it with the accepted value for your latitude and elevation above sea level.

5. Summarize all the above results in a neat table.

6. Construct a graph plotting T^2 as ordinates and L as abscissas, for Trials 2 through 5. Draw the ''best fit'' straight line through the four plotted points.

7. On another sheet of graph paper plot values of T (ordinates) against values of L (abscissas). Draw a ''smooth'' curve through these four plotted points.

Label the graphs carefully, and write an interpretation of the meaning of each curve.

ANALYSIS AND INTERPRETATION

1. Explain how the simple pendulum is an example of SHM. (This may require some additional reference reading on SHM.)

2. What do the results of Trials 6, 7, and 8, when compared to the result from Trial 4, reveal about the assumption which is inherent in the derivation of the formula for the period of a pendulum [Eq. (12.8)]?

3. Exactly how long ($L = $ length) must a pendulum be to beat seconds at a place where $g = 9.81$ m/s^{-2}?

4. Write a paragraph on the history of the discovery of the pendulum and of the use to which the discovery has been put in the development of time-keeping devices.

5. What is a Foucault pendulum? On what basic (Newtonian) principle does it operate, and what can be proved by observations on such a pendulum for a 24-h period?

Part 2. The Spiral Spring

A spiral spring and a weight holder are attached to a heavy laboratory stand as shown in Fig. 12.5, or a Jolly's balance can be used (see Fig. 12.6). A metric scale attached to the laboratory stand is helpful in reading displacements.

Figure 12.6. Jolly's balance. (*Central Scientific Co.*)

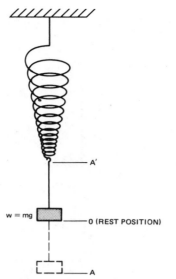

Figure 12.7. Rest position and amplitude in SHM with a spiral spring.

In Fig. 12.7 let O mark the "rest" or equilibrium position when $w = mg$ is the load. If the mass m is now pulled down to point A and released it will oscillate with a definite period T between A and A'. The distance $OA = OA'$ is called the *amplitude* of the motion.

The spring is assumed to be perfectly elastic, and therefore Hooke's law of elasticity applies. According to Hooke's law, the force F required to stretch a spring a distance x is

$$F = -kx \qquad (12.9)$$

where k is the force constant of the particular spring. The negative sign merely indicates that the direction of the force is opposite to that of the displacement. Actually, k is the force required to produce unit stretch of the spring. When the spring carries a load m and is displaced (stretched) from the equilibrium position a distance x, the restoring force $-kx$ is not exactly equal to ma, but to a quantity $m'a$, where m' includes not only the load mass m (including the weight holder or pan), but part of the distributed mass of the spring, as well. Therefore $-kx = m'a$, and

$$\frac{-x}{a} = \frac{m'}{k} \qquad (12.10)$$

Substituting from Eq. (12.10) in Eq. (12.6) gives the period of vibration of a spiral spring as

$$T = 2\pi \sqrt{\frac{m'}{k}} \qquad (12.11)$$

It must be recalled that k is the *force* constant of the spring, and in the SI-metric system the units of k will be in newtons per meter. The effective mass m' is the aggregate of the load mass m (including the hanger or pan) and some fraction f of the mass of the spring, m_s. Consequently,

$$m' = m + fm_s \qquad (12.12)$$

For laboratory-quality helical springs, including those

used on Jolly balances, the fraction f may be taken as one-third, as a close approximation.

MEASUREMENTS

(*Note:* Although measurements are, for convenience, taken in grams and centimeters, all data must be converted to *mks* units before calculations are carried out.)

Set up the apparatus as shown in Fig. 12.5 (or use Jolly's balance Fig. 12.6). Clamp a meter stick in a vertical position, so amplitudes can be observed. Determine the total load necessary to stretch the spring to a length about equal to twice its unloaded length. This should be the maximum load used during the experiment. The following suggested loads assume that this maximum load is 100 g. You should adjust the loads you use to fit the characteristics of the spring supplied with the apparatus.

1. Determining the Spring Constant Record a zero reading with only the weight holder (or pan) loading the spring. Then add successive loads in equal increments of about 10 g (depending on the stiffness of the spring) up to the previously determined maximum. There should be at least eight load increments. Record the values of the loads, scale readings, and the extension caused by each load. Convert loads (forces) into *mks* units.

2. Spiral Spring in Simple Harmonic Motion If the maximum load for your spring is about 100 g, use 25 g for an initial trial and set the system into vertical oscillation with an amplitude of about 5 cm. Measure the time for 50 *complete* oscillations, to the nearest 0.1 s. Repeat, as a check on the first run.

Then make two determinations of the time for 50 complete oscillations for loads (masses) of 50, 75, and 100 g (or loads which suit the stiffness of your spring). Express the load masses in *mks* units.

3. Determining the Mass of the Spring Determine m_s to the nearest 0.01 g on a laboratory balance.

CALCULATIONS

1. Plot a curve from the data in Step 1 above, to determine the spring constant, k. Plot loads in newtons vertically, and extensions of the spring horizontally. Draw the "best" straight line which approximates the locus of the plotted points. Determine the *slope* of this line by dividing the total ordinate (load) by the total abscissa (extension). The slope thus determined is the value of the spring constant k, in newtons/meter.

2. For the four determinations made in Step 2 under **Measurements,** calculate and tabulate the periods (in sec) and the squares of the periods. Plot a curve using the squares of the periods as ordinates and the *masses* used (kg) as abscissas. Draw the "best" line which approximates the locus of the plotted points. Extrapolate the line beyond the actual data until it crosses both the vertical (T^2) axis and the horizontal (mass-load) axis. From the intercept on the mass-load axis determine fm_s, that fraction of the mass of the spring which takes part in the SHM. Compare the value of fm_s taken from the graph, with the suggested value $1/3\ m_s$, which is typical for laboratory-type helical springs.

From Eq. (12.12) calculate m' (in kg) for each trial, and then compute T from Eq. (12.11) for each of the four load masses used. For each of the four trials, compute the percent difference between the actual observed value of T and the theoretical value obtained from Eq. (12.11).

Summarize all results in a neat table.

ANALYSIS AND INTERPRETATION

1. Explain how the period of a vertically oscillating spiral spring varies with: (1) the mass of the load; (2) the force constant of the spring; (3) the amplitude of the vibration.

2. Using dimensional analysis, prove that Eq. (12.11) is dimensionally correct—that the units of both sides are seconds.

3. Explain why the amplitude (within reasonable limits) is not a factor in the period for a loaded vibrating spring.

4. Why is it that the *period* of a loaded vibrating spring does depend on the load mass, but the period of a pendulum does not depend on the mass of the pendulum bob?

5. Show how you could use a vertically oscillating loaded spring to make an experimental determination of g, the acceleration of gravity.

NOTES, CALCULATIONS, OR SKETCHES

DATA SHEET

EXPERIMENT 12 ■ SIMPLE HARMONIC MOTION—THE PENDULUM AND THE SPIRAL SPRING

Part 1. The Pendulum

Trial	Length of pendulum (cm)	Pendulum bob material	Displacement (degrees)	Time for 50 complete oscillations (seconds)
1	200	Wood	3	
2	200	Metal	3	
3	150	Metal	3	
4	100	Metal	3	
5	50	Metal	3	
6	100	Metal	10	
7	100	Metal	20	
8	100	Metal	40	

Part 2. The Spiral Spring

1. The Spring Constant

Trial	Mass of load (including holder), (g)	Convert load (force) to newtons	Scale reading (cm)	Spring Extension (m)
1	Zero	Zero		Zero
2				
3				
4				
5				
6				
7				
8				
9				

2. The Spring in SHM

Trial	Mass of load (including holder) (kg)	Time for 50 oscillations (seconds)	
1		1st	
		2d	
2		1st	
		2d	
3		1st	
		2d	
4		1st	
		2d	

3. Mass of spring _____ g _____ kg

MOMENT OF INERTIA

PURPOSE

To study the principles of rotational dynamics and to determine by experiment the moment of inertia of several regularly shaped bodies about a vertical axis.

APPARATUS

Moment of inertia apparatus; laboratory rods, clamps, pulleys, string, weight holder, and slotted weights; solid metal disk, metal ring, cylindrical rod, or any other objects whose moment of inertia is to be determined; meter stick, stop clock, laboratory balance, vernier caliper, small spirit level.

INTRODUCTION

A system will be set in rotation about a vertical axis, with angular acceleration produced by a torque which is generated by a constant force supplied by a mass falling under the influence of gravity. It is assumed that, during the motion of the entire system, its *total energy* will be conserved, and the derivation of working equations is based on the principle of work or conservation of energy.

The apparatus (Figs. 13.1 and 13.2) consists of a turntable in the form of a lightweight aluminum cross, mounted in ball bearings, and rotating in a horizontal plane. Specimens whose moments of inertia are to be de-

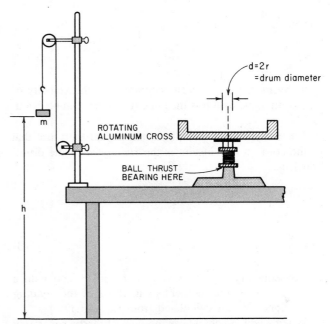

Figure 13.2. Diagram of the laboratory setup to study moment of inertia.

termined are placed on the turntable. The table is rotated with angular acceleration by means of the force of gravity acting on a mass *m*, falling through a vertical distance *h*. The falling mass is connected by a cord to a drum which is a part of the rotating turntable, and by which the torque is applied.

Derivation of Equations   Let r be the radius of the drum (Fig. 13.2) around which the cord is wound; h the distance through which a mass m descends, being accelerated by the force of gravity g; v its velocity as it hits the table top (or floor); t the observed time of fall in seconds; I the moment of inertia of the rotating system; and ω the angular velocity of the rotating system that has been attained at the instant m hits the table top or floor.

From conservation of energy principles, the initial *potential energy* of the system will be converted into *translational kinetic energy* of the falling mass *plus* the *rotational kinetic energy* of the rotating system. We can write

$$mgh = \tfrac{1}{2}mv^2 + \tfrac{1}{2}I\omega^2 \tag{13.1}$$

where I is the *moment of inertia* of the rotating system.

The driving force is constant and therefore the linear acceleration will be uniform, and will be denoted by a.

Figure 13.1. Rotating aluminum cross for moment of inertia experiment. Objects whose moments of inertia are to be experimentally determined are placed on the rotating cross, whose moment of inertia (alone) is first determined. *(Central Scientific Co.)*

Since, from Newton's laws of accelerated motion (for translation),

$$h = \tfrac{1}{2}at^2$$

then

$$a = \frac{2h}{t^2}$$

and since

$$v = at$$

we obtain

$$v = \frac{2h}{t} \qquad (13.2)$$

Thus by measuring h and t accurately, v, the velocity of the falling mass as it hits the table (or floor), can be calculated.

But the speed of the falling mass (and therefore that of the cord) at any instant is the *rim speed* of the drum, which is

$$v = r\omega$$

from which

$$\omega = \frac{v}{r} \qquad (13.3)$$

Consequently, by measuring h, t, and r, and with a known mass m, the moment of inertia I of the *rotating cross* alone may be calculated from Eq. (13.1).

When a new object whose moment of inertia separately is I_1 is placed on the rotating cross, and a mass m_1 is used to supply the turning force, the moment of inertia of the *new object* may be obtained from

$$m_1 g h_1 = \tfrac{1}{2} m_1 v_1{}^2 + \tfrac{1}{2}(I + I_1)\omega_1{}^2 \qquad (13.4)$$

where I is the moment of inertia of the cross alone.

MEASUREMENTS

Note: The SI-metric system only will be used and *mks* units will be employed in the calculations.

Measure the diameter of the drum with the vernier caliper and compute the radius r. Take several trials and average the results.

Now level the rotating system carefully and set up the rod, pulley, and cord system. Attach small "riders" to a loop in the cord in ½ g steps until the cross just rotates with a constant speed when started by a gentle push. These riders compensate for friction in the apparatus, and are small in magnitude in comparison with the falling mass which accelerates the system.

Note: The mass of the friction riders (m_r) *should not be included* in the value of m as used on the *left sides* of Eqs. (13.1) and (13.4), but *should be included,* in the interest of rigor, as a part of the m's on the *right side* of these equations. (Why?)

Hang mass m (a magnitude of 20 to 70 g is suggested)

on the cord and mark off a carefully measured length h (2 m is a good distance). Make four accurate determinations of the time of fall t with no load on the rotating cross. Record all data on the Data Sheet.

Now place the flat disk on the rotating cross, add the necessary friction riders, and then mass m_1 (perhaps 100 to 300 g). Make four accurate determinations of h_1 and t_1.

Repeat with the cylindrical ring, the solid cylindrical rod, and any other objects your instructor may provide. For each object first determine the mass of the friction rider(s) required, and use these values of m_r as noted above.

Measure accurately all the linear dimensions and the masses of the objects for which I is to be determined.

CALCULATIONS

1. From Eq. (13.1), calculate I for the system with the aluminum cross alone.
2. From Eq. (13.4), calculate I_1 for each of the bodies placed on the rotating system.
3. Calculate the moments of inertia of all bodies used from the formulas given in Fig. 13.3, and compare these calculated values with the experimental results you ob-

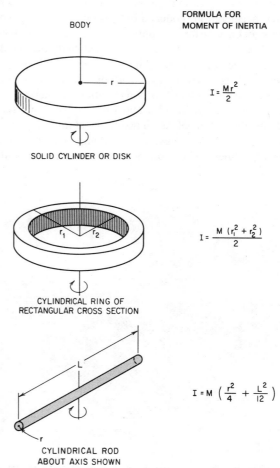

BODY	FORMULA FOR MOMENT OF INERTIA
SOLID CYLINDER OR DISK	$I = \dfrac{Mr^2}{2}$
CYLINDRICAL RING OF RECTANGULAR CROSS SECTION	$I = \dfrac{M(r_1^2 + r_2^2)}{2}$
CYLINDRICAL ROD ABOUT AXIS SHOWN	$I = M\left(\dfrac{r^2}{4} + \dfrac{L^2}{12}\right)$

Figure 13.3. Formulas for the moments of inertia of three regularly shaped bodies. See any standard physics text for others, or consult your instructor.

tained. Compute the percent error of your experimental results.

ANALYSIS AND INTERPRETATION

1. Discuss the precision, or lack of it, of your measurements of m, t, and h. If you made a 1 percent error in measuring the time t, what error would result in your calculation for the total moment of inertia $(I + I_1)$ in Eq. (13.4)?

2. If the laboratory has a rotating table on which you can stand, experiment with the effects of moment of inertia by extending your arms (with a weight in each hand) while someone rotates you slowly. Then bring your arms and the weights smartly in close to your body and notice the result. Be careful until you have practiced a bit! Explain what you observe.

3. This experiment is based on conservation of energy principles, and Eqs. (13.1) and (13.4) are expressions of conservation of energy. Is there some other basis on which the experiment could be performed, *with the same apparatus*, to determine I? (*Hint:* Note that the very fundamental relationship $\tau = I\alpha$ was not used in the discussion or the calculations with the *conservation of energy* method.) Briefly describe the essential steps in an experiment to determine I from torque-and-acceleration considerations.

4. Explain why large flywheels are cast with most of the mass in an outer rim and with the spokes having as little mass as possible.

5. Discuss the relationship of moment of inertia and angular momentum. What applications do these concepts have to such sports events as figure skating, high diving, gymnastics, and discus throwing?

DATA SHEET

EXPERIMENT 13 ■ MOMENT OF INERTIA

Rotating system	Trial	Falling mass, m (kg)	Friction rider(s), m_r (kg)	Height h (m)	Time, t (seconds)
Empty cross	1				
	2				
	3				
	4				
Cross with solid cylinder or disk	1				
	2				
	3				
	4				
Cross with cylindrical ring	1				
	2				
	3				
	4				
Cross with _____	1				
	2				
	3				
	4				

Record below the dimensions and masses of the regularly shaped bodies used:

PART TWO ■ PROPERTIES OF MATTER

EXPERIMENT 14

SOME PROPERTIES OF SOLIDS—ELASTICITY AND TENSILE STRENGTH

PURPOSE

To study the elasticity and tensile strength of a steel wire, and to determine Young's modulus and the ultimate tensile strength of steel.

APPARATUS

Young's modulus apparatus (sometimes called Searle's apparatus); pound weights (or kilogram masses) and weight hanger; drawn steel and brass test wires; micrometer calipers; ultimate tensile-strength apparatus.

Note: In lieu of the Searle's apparatus, a heavy duty laboratory stand may be used, with a suitable cathetometer.

INTRODUCTION

Every substance is elastic to some degree. A material is said to be elastic if after a deformation of any kind, it returns readily to its original shape. The popular notion is that rubber bands, tennis balls, and golf balls are good examples of elasticity. They are somewhat elastic to be sure, but many materials, including steel and ivory, are much more elastic than rubber.

There is a limit to the force which may be applied to a body and still have it return to its original, undistorted state. This is called the *elastic limit*. If the elastic limit of the material is exceeded, it acquires a *permanent set*. If a stretching force is increased beyond the elastic limit, the material will eventually pull apart or break.

If a wire is held fast at one end and a small force ΔF is applied at the other, the force applied divided by the cross-section area of the wire is called the *tensile stress* (see Fig. 14.1).

$$\text{Tensile stress} = \frac{\text{applied force}}{\text{area}} = \frac{\Delta F}{A} \qquad (14.1)$$

This applied force ΔF will stretch or elongate the wire a small amount. If the original length of the wire is L, and the small elongation is ΔL, the ratio $\Delta L/L$ is called the *tensile strain*. (Fig. 14.1).

$$\text{Tensile strain} = \frac{\text{increase in length}}{\text{original length}} = \frac{\Delta L}{L} \qquad (14.2)$$

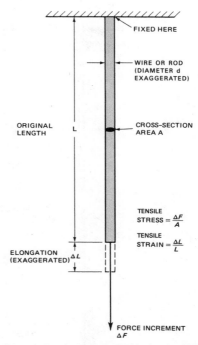

Figure 14.1. Diagram showing the relationships involved in Young's modulus of elasticity using a stretched wire. The elongation (amount of stretch) is grossly exaggerated, as is the wire diameter.

Hooke's Law Robert Hooke conducted many experiments on elasticity and discovered the relationship between stress and strain. *Hooke's law of elasticity* states that

For elastic materials within their elastic limit, strain is directly proportional to stress.

Young's Modulus The ratio of stress to strain is a constant characteristic of a given material, independent of the shape or size of the specimen under test. This ratio, *for stretching or compressional stresses and strains,* is called Young's modulus, Y.

$$Y = \frac{\text{tensile stress}}{\text{tensile strain}} = \frac{\Delta F/A}{\Delta L/L} = \frac{L\Delta F}{A\Delta L} \qquad (14.3)$$

Some Properties of Solids—Elasticity and Tensile Strength **79**

Ultimate Tensile Strength The force per unit area (stress) at the instant a material pulls apart is known as the *ultimate tensile strength*.

Note: This experiment is written for English-system units, but mks units may be used if your instructor prefers.

MEASUREMENTS

Part 1. Young's Modulus of Elasticity

Secure the steel test wire firmly at the top chuck of the Searle's apparatus. (Fig. 14.2.) Pass the other end through the bottom chuck, between the vertical support rods. Form a loop at the bottom in which a weight hanger may be

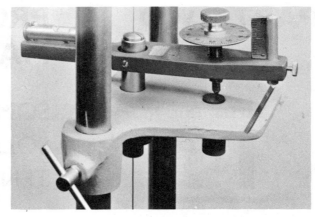

Figure 14.3. Close-up of Searle's apparatus, showing micrometer screw, three-jaw chuck for gripping test wire, and the spirit level. *(Central Scientific Co.)*

placed (see Fig. 14.2.[1]) Put on a 2-lb weight to draw the wire taut, and level the laboratory stand so weights hang free and vertical. Then tighten the bottom chuck on the wire and level the Searle's apparatus with the micrometer screw (see Fig. 14.3). Record this as the "zero reading." Add weights 2 lb at a time, releveling and reading the micrometer each time.[2] Record readings in the Data Table. (For a steel wire of about 0.030 in. in diameter, a maximum load of 20 lb is satisfactory.) When the maximum load has been reached, take the weights off one at a time in reverse order, leveling the apparatus and reading the micrometer each time. Record these readings.

Measure the diameter of thē wire with a micrometer caliper to the nearest 0.001 in. (Take several measurements and average the results.) Measure the length of the wire between chucks, accurate to 0.05 in.

Repeat all the above with a brass test wire, and with others your instructor may supply.

Part 2. Ultimate Tensile Strength

Observe the ultimate tensile strength demonstration (Fig. 14.4) performed by the instructor, and record the data needed to calculate the ultimate tensile strength of the test sample. *Stand clear at all times and wear goggles for eye protection!*

CALCULATIONS

1. Convert all readings to inches and pounds.[3] Calculate the average elongation, in inches, produced by a stretching force of 2 lb. Calculate the cross-section area *A* of the

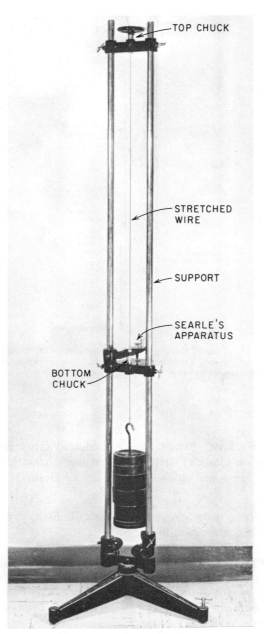

Figure 14.2. Laboratory support stand and Searle's apparatus setup for the determination of Young's modulus by stretching a metal wire. *(Central Scientific Co.)*

---

[1] If a cathetometer is to be used instead of Searle's apparatus, glue a paper marker on the wire just above the weight hanger, and set up the cathetometer on a firm support so that it can travel in the direction the wire will be stretched.

[2] If a cathetometer is being used, run the cross hair to the paper marker you placed on the wire, and read the cathetometer vernier scale.

[3] Engineering and construction materials data are still expressed, for the most part, in English units in the United States. Use mks units, however, if your instructor prefers. Remember the mks-system relation between force and mass!

The figure 14.2 labels: TOP CHUCK, STRETCHED WIRE, SUPPORT, SEARLE'S APPARATUS, BOTTOM CHUCK

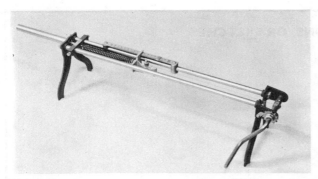

Figure 14.4. Apparatus for determining the ultimate tensile strength of a test wire. *(Central Scientific Co.)*

wire, and then compute Young's modulus for steel and for brass from Eq. (14.3).

2. Compute the ultimate tensile strength of the test wire from the demonstration results. Compare your results with accepted values from an engineering handbook, and calculate the percent error of your results.

Put all your calculated results in a neat summary table.

ANALYSIS AND INTERPRETATION

Discuss the sources of error in this experiment. Write a page or more on the applications of a knowledge of Young's modulus and ultimate tensile strength to industry, engineering, construction, and even to musical instruments like the piano. See if you can find, in tables, Young's modulus and the ultimate tensile strength of piano wire.

To demonstrate your knowledge of percent error, estimate roughly, *without pencil or calculator,* the approximate percent error involved in the following:

1. A laboratory experiment gives 174,000 mi/s as the speed of light. The accepted value is 186,000 mi/s.

2. A precision part for a jet engine is supposed to be 2.056 in. in diameter. A sample is measured and found to be 2.052 in. in diameter.

3. An experiment gives 7.65 g/cm^3 for the density of drawn steel. The handbook value is 7.84 g/cm^3.

Explain the steps in your reasoning in arriving at each of the estimated percent errors.

Problem A cable of drawn steel is used with a crane to lift junked automobiles. (a) If the stress in the cable while supporting an automobile with a weight of 3000 lb is not to exceed one-half the ultimate tensile strength of the cable, what must be the diameter of the cable? (b) If the unloaded cable has a length of 20 ft and a diameter determined by the above guideline, how much will the cable stretch when it is used to support the 3000-lb automobile? (Assume the ultimate tensile strength for steel is 7.8 × 10^4 lb/in.^{2} and that Young's modulus for steel is 28 × 10^6 lb/in.^{2}.)

NOTES, CALCULATIONS, OR SKETCHES

Name _____ Date _____ Section _____

Course _____ Instructor _____

DATA SHEET

EXPERIMENT 14 ■ SOME PROPERTIES OF SOLIDS—ELASTICITY AND TENSILE STRENGTH

Part 1. Young's Modulus

Load (lb) (see note below.)	Steel wire micrometer readings (or cathetometer readings)		Brass wire micrometer readings (or cathetometer readings)	
	Loading	*Unloading*	*Loading*	*Unloading*
"Zero reading"				
2				
4				
6				
8				
10				
12				
14				
16				
18				
20				

	Diameter (ave. of 3 trials)	Length (between chucks)
Steel wire	_____ in.	_____ in.
Brass wire	_____ in.	_____ in.

Part 2. Ultimate Tensile Strength

Diameter of test wire _____ in. (ave. of three trials)

Breaking force _____ lb

Wire material _____

Note: Mks units may be used if instructor prefers. In that case a maximum of 10 kg is suggested.

NOTES, CALCULATIONS, OR SKETCHES

EXPERIMENT 15

TORQUE AND SHEAR MODULUS

PURPOSE

To study the effects of torque applied to cylindrical rods, and to determine the shear modulus (coefficient of rigidity) of mild steel, brass, and copper.

APPARATUS

Torsion apparatus, designed to clamp a cylindrical rod at one end and apply a measurable torque at the other; masses to be used as loads, micrometer caliper, and meter stick; cylindrical test rods of steel, brass, and copper, of various diameters and lengths.

INTRODUCTION

The English physicist Robert Hooke demonstrated that the *linear* deformation of an elastic body is proportional to the applied stretching or compressional force, if the *elastic limit* is not exceeded (see Experiment 14). In like manner, if a force is applied to a cylindrical rod in such a manner as to produce a *twisting effect*, or *torque*, on the rod, the stain in the rod is found to be a *shearing* strain, and the amount of strain (angle of twist) is proportional to the torque applied. For a given material, the length of the rod and its diameter naturally affect the amount of twist produced by a given torque. The exact relationship between the torque applied and the angle of twist is not a simple one and it will be given without derivation, since the analysis involves calculus-level mathematics.

If, in Fig. 15.1, $\tau = FR$ is the twisting torque, θ the angle of twist produced (radians), L the length of the rod, and r the radius of the rod, then

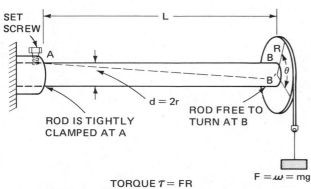

SET SCREW

A

ROD IS TIGHTLY CLAMPED AT A

$d = 2r$

ROD FREE TO TURN AT B

$F = w = mg$

TORQUE $\tau = FR$
ROD IS TWISTED SO THAT ELEMENT AB ASSUMES THE LINE OF AB′. $\theta = $ ANGLE OF TWIST.

Figure 15.1. Torque applied to a long cylindrical rod—a diagram of the factors involved in determining shear modulus.

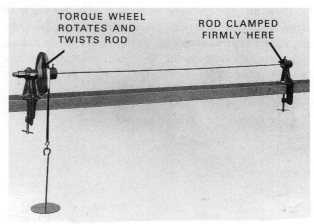

TORQUE WHEEL ROTATES AND TWISTS ROD

ROD CLAMPED FIRMLY HERE

Figure 15.2. Apparatus assembled for applying torque to a cylindrical rod to determine the shear modulus of the material. *(Central Scientific Co.)*

$$\theta = \frac{2\tau L}{\pi s\, r^4} \qquad (15.1)$$

and,

$$s = \frac{2\tau L}{\pi \theta r^4} \qquad (15.2)$$

In Eqs. 15.1, and 15.2, s is the *shear modulus*, or *coefficient of rigidity*, of the material. It is a constant (within the elastic limit) for a given material. In the *mks* system, s is in newtons/m² when τ is in newton-meters, L and r are in meters, and θ is in radians.

MEASUREMENTS

Insert and center the short, small-diameter steel rod in the torsion apparatus, clamping it tightly into both ends of the equipment. Be sure the entire apparatus is securely clamped to the laboratory table (see Fig. 15.2). Attach a weight holder to the end of the strap fitted to the torque wheel, and load it sufficiently to take the slack out of the apparatus. This becomes the "zero load." Set the angular measurement vernier scale on zero. Test for nonslippage by adding a load which will twist the rod about 60° and then remove this load. The wheel should return to the zero of the vernier scale. If it does not, determine the cause of slippage and correct it before proceeding.

Now place a 200-g load on top of the "zero load" and read and record the angle of twist from the scale and vernier. Continue readings using steps of 200 g until total twist of about 90° is reached.

Determine the radius R of the torque wheel, the radius r of the test rod (micrometer caliper), and the length of the rod.

Repeat with a longer steel rod of the same diameter; then again with a steel rod of a larger diameter. Repeat with rods of different materials, such as copper and brass.

Record all data on the Data Sheet.

CALCULATIONS

1. Show that for the same material (steel), the angle of twist varies directly with the twisting torque and the length, but inversely with the fourth power of the radius.

2. Express all units in the SI-metric (mks) system and calculate s for each material, from each trial. Do the results verify Hooke's law?

3. Compare your results for shear modulus to accepted values for s for the materials you used (see tables in the Appendix), and calculate the percent error of your results.

ANALYSIS AND INTERPRETATION

1. Discuss the relative effects on the accuracy of your results of the precision with which you were able to measure L, r, and θ.

2. If a 2-in. diameter pump shaft of mild steel goes 830 ft deep into a well, and if the steady torque being applied by the electric motor at the top is 300 lb · ft, find the total angle of twist of the shaft. [Use English units in Eq. (15.1).]. What are the English-system units of s?

3. Work has to be done to twist a shaft. What happens to this work as the shaft undergoes shear and twists through an appreciable angle?

DATA SHEET

EXPERIMENT 15 ■ TORQUE AND SHEAR MODULUS

NOTE: Only a skeleton data sheet is shown. The student is to work out the rest of it. Note that mks units are used.

Radius R of torque wheel _____m

Type of rod	Trial	Load (kg)	Total twist (rad)	Twist per 0.2-kg load (rad)
Short steel, small diameter $L =$ _____ m $r =$ _____ m	1 2 3 4 5 6 7 8 etc.	0.200 0.400 0.600 0.800 1.00 1.20 1.40 1.60 etc.	 (max. = 90°)	
Long steel, small diameter $L =$ _____ m $r =$ _____ m	1 2 3 4 5 6 7 etc.			

This format may be continued for as many types of rods as your instructor wants you to test.

NOTES, CALCULATIONS, OR SKETCHES

EXPERIMENT 16

ARCHIMEDES' PRINCIPLE—
BUOYANCY AND FLOTATION

PURPOSE

To study the principles of buoyancy and flotation, and to verify Archimedes' principle for both floating and submerged bodies.

APPARATUS

Overflow can and catch bucket, platform balance, graduated cylinder, heavy object and light and light object, hydrometer for "heavy" and "light" liquids, thermometer, thread.

INTRODUCTION

Archimedes' principle states that *an object floating on or immersed in a fluid is buoyed up by a force equal to the weight of the fluid displaced*. If the object is floating, the buoyant force is obviously equal to its weight; therefore, a floating object is said to displace its own *weight* of the fluid in which it floats (see Fig. 16.1). If, on the other hand, the (solid) object is more dense than the fluid and sinks, it displaces its own *volume* of the fluid, and the buoyant force upward on the object is equal to the weight of a volume of the fluid which is equal to the volume of the object (see Fig. 16.2).

Let w be the weight of an object in air and w_1 its *apparent* weight when submerged in a liquid. [The buoyant ef-

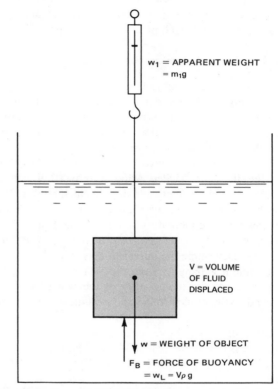

Figure 16.2. *Immersed object.* The object is more dense than the fluid in which it is placed, and therefore it sinks. The net downward force (the *apparent weight*) is $w_1 = w - F_B$. A solid object displaces its own volume of a fluid in which it sinks. This volume V, multiplied by the weight density of the fluid ρg, equals the weight of the fluid displaced, w_L. And, from Archimedes' principle, the buoyant force equals the weight of the fluid displaced, or $F_B = w_L$.

fect of the air (which is a fluid) can be neglected in this experiment without causing appreciable error.] Then

Buoyant force = apparent loss in weight

$$F_B = w - w_1 = mg - m_1g \tag{16.1}$$

Where m is the actual mass of the object and m, is its *apparent mass* when submerged in a liquid. Let the volume of the liquid displaced be designated by V, and its mass density[1] by ρ. Then the weight of liquid displaced, $w_L = V\rho g$, from which, by Archimedes' principle,

$$F_B = V\rho g \tag{16.2}$$

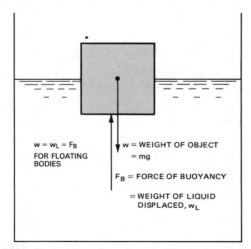

Figure 16.1. *Floating object.* Since the object is in vertical equilibrium, w (the actual weight of the object) = F_B (the upward force of buoyancy on the object). But since F_B = the weight of the displaced fluid (from Archimedes' principle), a floating object must displace its own weight of the fluid in which it floats.

---

[1]*Mass density* is defined as the mass per unit volume of a substance. *Weight density* (D) is *weight* per unit volume. Mass density (ρ) is related to weight density D by $D = \rho g$, where g is the acceleration of earth's gravity.

In terms of Eqs. (16.1) and (16.2), Archimedes' principle says that

Buoyant force on the object = apparent loss of weight of the object = weight of displaced liquid

This statement is true whether the body is floating or submerged.

MEASUREMENTS

The SI-metric system will be used in this experiment. From the following measurements, Eqs. (16.1) and (16.2) can be verified and Archimedes' principle proved.

Part 1. Using Water as the Liquid

Suspend the "heavy" object from a platform balance as shown in Fig. 16.3, and determine its mass in air to the nearest 0.1 g. This is the mass m in Eq. (16.1). The object should be *quite large*, so that it will displace a considerable volume of liquid, and quite heavy.

Fill the overflow can with water until it overflows freely into the catch bucket. Allow it to overflow *until all dripping stops*, then empty and dry the catch bucket. Now bring the balance to the overflow can and lower the balance slowly on the laboratory stand until the heavy object has sunk below the water surface. Catch all the water that overflows. Determine the *apparent* mass m_1 of the heavy object while it is hanging from the balance submerged in water. Measure the temperature of the water. With the graduated cylinder determine the volume of water displaced.

Repeat all of the above measurements with an object which is less dense than water—a floating object. Its *apparent mass* m_1 will of course be zero. Enter all data in the table.

Part 2. Using a Liquid Heavier (More Dense) than Water

Repeat all the measurements of Part 1 (except temperature of the liquid) for both objects, using a liquid heavier than water. Measure the density ρ of the liquid with the hydrometer.

Part 3. Using a Liquid Lighter than Water

Repeat all measurements for both objects using a liquid lighter than water (such as methyl alcohol). Measure the density ρ of the liquid with the hydrometer.

CALCULATIONS

1. From a handbook determine the density (g/cm³) of water at the temperature you recorded.
2. For all three liquids, show that you have verified Archimedes' principle for *floating* bodies; i.e., a floating body displaces its own weight of the fluid in which it floats. *Note:* Display your results in such a way that there will be

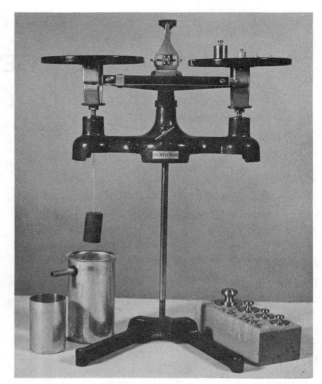

Figure 16.3. Method of suspending the object from a platform balance to determine its mass in air (m) and its *apparent mass* (m_1) when immersed in the liquid. The arrangement of overflow can and catch bucket is also shown.

no doubt about the proof of the principle. This will call for some critical thinking about the nature of proof.
3. For both the heavy and light liquids, and for water, show that you have verified Archimedes' principle for *submerged* bodies, by use of Eq. (16.2).

ANALYSIS AND INTERPRETATION

Analyze your errors and discuss sources of error. Write a paragraph on each of the following:

1. The history of this experiment—Archimedes' original discovery.
2. The practical applications of buoyancy and flotation. Include applications to ships, submarines, balloons, and industrial flotation processes.

Solve the following problems:

1. An object whose mass is 275 g (in air) has an apparent mass of 185 g when suspended in a heavy liquid. If the volume of liquid displaced is 75 cm³, find the density of the object and the density of the liquid.
2. At sea level a balloon when inflated occupies 200,000 ft³. It is filled with helium at atmospheric pressure (air temperature 70°F). The weight of fabric and gondola combined is 6000 lb. Find the net lifting force in pounds. (See handbook tables for densities of air and helium at 70°F.) In the English system density (D) is expressed in lb/ft³.

Name _____ Date _____ Section _____

Course _____ Instructor _____

DATA SHEET

EXPERIMENT 16 ■ ARCHIMEDES' PRINICIPLE—BUOYANCY AND FLOTATION

Liquid	Object	Mass m of object in air (g)	Apparent mass m_1 of object in liquid (g)	Volume V of overflow liquid (cm^3)	Mass of catch bucket and overflow water (g)	Temperature of liquid (°C)
Water	Heavy object					
	Floating object					
Liquid heavier than water	Heavy object					Density of liquid, from hydrometer, _____ g/cm^3
	Floating object					
Liquid lighter than water	Heavy object					Density of liquid _____ g/cm^3
	Floating object					

Mass of empty catch bucket _____g

NOTES, CALCULATIONS, OR SKETCHES

EXPERIMENT 17

BOYLE'S LAW OF GASES

PURPOSE

To study the relationship between pressure and volume of a given mass of gas when the temperature is constant, and to verify Boyle's law of gases.

APPARATUS

Boyle's law apparatus; laboratory barometer.

INTRODUCTION

Unlike solids and liquids, gases have no definite volume. The volume of a given mass of gas varies markedly with changes in pressure and temperature. Pressure, volume, and temperature relationships in gases actually determine the *condition* of the gas at any time. We refer to these variables as the *P-V-T* conditions of a gas.

In actual situations it is quite probable that pressure and temperature will vary simultaneously, but for purposes of laboratory study it is possible to hold either pressure or temperature constant and observe the effect of a change in the other on the volume of a given mass of gas. In this experiment, temperature will be held constant (room temperature) and the effect of a variation of pressure on the volume of a gas will be observed.

Robert Boyle first studied this relationship in 1662, and his findings have since been known as Boyle's law. Boyle found that, if the temperature of a *given mass* of gas is held constant while the *absolute pressure* is varied, the volume of the gas varies inversely with the absolute pressure. This relationship can be expressed as an equation:

$$\frac{V_1}{V_2} = \frac{P_2}{P_1} \tag{17.1}$$

or

$$P_1V_1 = P_2V_2 = P_3V_3, \text{ etc.}$$

Since any number of *PV* conditions may apply to a given mass of gas, it is apparent that the *product* of absolute pressure times volume is a constant, or

$$PV = k \tag{17.2}$$

This equation is the accepted mathematical form of Boyle's law. Mathematically, Eq. (17.2) is the equation of a hyperbola, and if values of absolute pressure are plotted as ordinates and values of corresponding volumes as *abscissas*, an equilateral hyperbola, as shown in Fig. A in the Introduction will be obtained. The student should note that in Fig. A the area of rectangle P_1AV_1O (length times

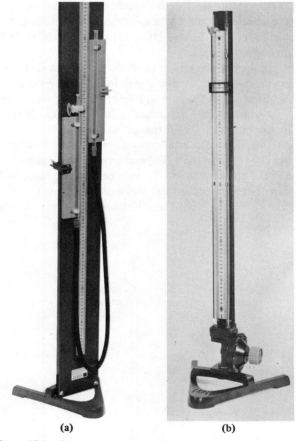

Figure 17.1. Two forms of apparatus for verifying Boyle's law. (*a*) Elevation pressure-head form, described in the experiment instructions; (*b*) mercury-well, diaphragm pressure form. *(Central Scientific Co.)*

width is P_1V_1) is equal to that of P_2BV_2O (length times width is P_2V_2).

The Boyle's law apparatus recommended (see Fig. 17.1*a*) consists of two vertical glass tubes of uniform bore, one open at the top and the other either sealed off or fitted with a stopcock. The bottom ends of the tubes are connected by rubber or plastic tubing. Either glass tube may be raised or lowered and clamped in position. When the apparatus is filled with mercury, the mercury levels in the two tubes may be controlled at will, thus changing the pressure and the volume of the air trapped in the sealed tube on the left side of the apparatus.

The apparatus is fitted with a meter stick which facilitates reading the mercury levels. Changes in pressure should be made very slowly (by elevating or lowering the right-hand tube) in order to prevent temperature changes which might result from the work of compression. Wait at least a full minute after each mercury level change before taking a reading on the air volume trapped in the (left-hand) sealed tube. Handle the apparatus very carefully! *Mercury is poison!* If any is spilled, notify the instructor immediately. If any mercury contacts the skin, wash up at once.

MEASUREMENTS

The volume of the confined air in the left-hand tube, Fig. 17.2, (stopcock closed) should be about one-half the total volume of that tube when the mercury is at the *same level* in both tubes. At this setting the air in *both* tubes will be atmospheric pressure.

Preparatory to the taking of readings, test for leaks by raising the open tube as high as possible. Leave it there for a few minutes. If the mercury level in the closed tube remains constant, the apparatus is free of leaks. If not, report the matter to the instructor.

In terms of the lettered points on Fig. 17.2, a represents the mercury level in the right-hand tube and b the mercury level in the left-hand tube. The volume V of the entrapped gas (air) is expressed in arbitrary units (not cm^3) as the length $c - b$ of the tube. The differential pressure p (in cm

Hg) is $a - b$. The absolute pressure P is the sum of the differential pressure p and the barometric (atmospheric) pressure B. When a is lower than b, the pressure in the left-hand tube is *less than atmospheric*. Proper regard to algebraic signs must be observed in calculating p and P.

Now lower the right-hand column as far as the instrument will permit. (Do not spill the mercury!) Read levels a, b, and c from the meter stick scale. Then take a series of corresponding readings of levels a and b, increasing the differential pressure p in 5-cm steps by raising the open tube, until the maximum height (pressure) is attained. Make the changes slowly and wait one minute between readings. (Why?) Enter the data in the Data Table, and read the laboratory barometer (Fig. 17.3), to determine the atmospheric pressure at the time, in cm Hg.

CALCULATIONS

1. For each set of readings calculate the product PV. Do your calculations show that the product PV is a constant—that Boyle's law is verified?

2. Plot a "smoothed" curve on graph paper, using *absolute pressures* in cm Hg (P) as ordinates, and arbitrary units of volume (V) as abscissas. (Review the section on plotting and interpreting curves and graphs in the Introduction.) Prepare this graph with great care, since it is the essence of your report.

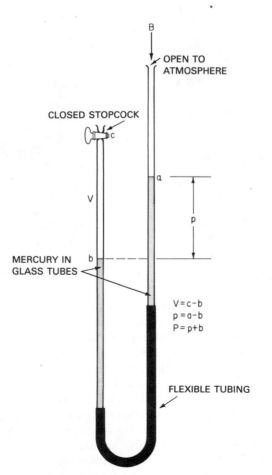

B

OPEN TO ATMOSPHERE

CLOSED STOPCOCK

c

V

a

p

MERCURY IN GLASS TUBES

b

$V = c - b$
$p = a - b$
$P = p + b$

FLEXIBLE TUBING

Figure 17.2. Diagram showing the relationship between volume, differential pressure, and absolute pressure as involved in the Boyle's law apparatus.

3. Also plot values of P (vertically) against corresponding values of $1/V$, the reciprocals of the volume readings. What is the significance of the resulting "best fit" line through the plotted points?

ANALYSIS AND INTERPRETATION

1. Analyze and discuss any significant discrepancies between your results and Boyle's law.

2. Explain in detail what effect a change in temperature during the experiment would have on the results.

3. At what extremes of temperature and pressure does Boyle's law no longer apply? Explain.

4. Devise and explain a simple and convenient method of converting pressures in cm Hg into corresponding values in lb/in.^{2}.

NOTES, CALCULATIONS, OR SKETCHES

Name _____ Date _____ Section _____

Course _____ Instructor _____

DATA SHEET

EXPERIMENT 17 ■ BOYLE'S LAW OF GASES

Trial	Scale reading at stopcock c (cm)	Scale reading at b (cm)	Scale reading at a (cm)	$V = c - b$ (arbitrary units)	$p = a - b$ (proper regard to algebraic sign) (cm)	$P = p + B$ (cm Hg)
1						
2						
3						
4						
5						
6						
etc.						

Barometer reading B _____ cm Hg

NOTES, CALCULATIONS, OR SKETCHES

PART THREE ■ HEAT AND THERMODYNAMICS

EXPERIMENT
18

TEMPERATURE AND THERMOMETERS— CALIBRATION OF A THERMOMETER

PURPOSE

To calibrate an unengraved thermometer, and to use that thermometer to measure the air temperature in the laboratory.

APPARATUS

Unengraved mercury thermometer; beaker; tripod support for beaker; burner; pen for marking glass; meter stick; Celsius and Fahrenheit laboratory reference thermometers, (accurate to 0.2 degree)

INTRODUCTION

A standard laboratory thermometer consists of a glass reservoir bulb and a very narrow tube or capillary partially filled with mercury, as shown in Fig. 18.1. The portion of the capillary tube above the mercury is evacuated so that the mercury can easily expand, upon heating, into the unfilled portion of the capillary. If the thermometer is moved to a location where the air is warmer, or if it is immersed in a hot liquid, both the glass and the mercury will expand. But the glass dimensions are approximately 16 times less sensitive to temperature changes than the mercury dimensions are. Moreover, the volume of the reservoir bulb and the diameter of the capillary are chosen to emphasize the change in volume of the mercury as observed in the capillary. The location of the end of the mercury column in the capillary is used, with the aid of an associated scale, to determine the temperature of the gas, liquid, or solid surrounding the thermometer.

Two common temperature scales are the *Fahrenheit* and the *Celsius* scales. On the Fahrenheit scale the freezing point of water is 32°F and the boiling point of water is 212°F. On the Celsius scale the freezing point of water is 0°C and the boiling point of water is 100°C. Since there are 180 Fahrenheit degrees and 100 Celsius degrees be-

tween the freezing and boiling points of water, a Celsius degree is 1.8 times as large as a Fahrenheit degree.

MEASUREMENTS

Fill a glass beaker two-thirds full of chipped (or cube) ice. Add just enough tap water to cover the ice. Let stand a few minutes in order that the lowest possible temperature will be attained. Now hold the unengraved thermometer in the ice-water mixture until the length of the mercury column stops decreasing. Use a pen with waterproof ink to make a *freezing-point mark* on the glass next to the end of the mercury column. Remove the thermometer and pour out the ice-water mixture.

Fill the beaker half full with water and use the burner to bring the water to a vigorous boil. Hold the end of the thermometer *in the boiling water* until the length of the mercury column stops increasing. Use the pen to make a *boiling-point mark* on the glass next to the end of the mercury column.

Hold the thermometer under running tap water for several seconds and then dry the thermometer with a paper towel. Place the thermometer on the laboratory bench and wait several minutes for the length of the mercury column to stop changing. Make a *room-temperature mark* on the glass next to the end of the mercury column. Your thermometer should now look something like the one in Fig. 18.2.

Measure accurately and record the distance d_1 between the freezing- and boiling-point marks. Measure and record the distance d_2 between the freezing-point mark and the room-temperature mark.

CALCULATIONS

If the mercury in the capillary expands by equal amounts in response to equal temperature increases throughout the range of temperatures from the freezing point to the boiling point, the room temperature will be the fraction d_2/d_1 of 100°C. Use your measured values for d_1 and d_2 and the following equation to calculate the room temperature T_R (C°).

$$T_R(\text{C}°) = (d_2/d_1)100 \qquad (18.1)$$

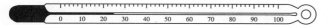

Figure 18.1. A typical laboratory mercury-in-glass thermometer, Celsius scale.

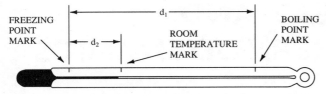

Figure 18.2. Unengraved thermometer at room temperature showing calibration marks.

Compare your calculated value for room temperature with the value obtained by reading the Celsius *laboratory reference thermometer*. Compute the percent difference between the two values.

For the Fahrenheit temperature scale there are 180 degrees between the freezing point (32°F) and the boiling point (212°F). Use your measured values for d_1 and d_2 and the following equation to calculate the room temperature in Fahrenheit degrees.

$$T_R(\text{F}°) = (d_2/d_1)\ 180 + 32 \qquad (18.2)$$

Compare this experimental value with the room temperature reading obtained with a Fahrenheit *reference thermometer*. Calculate the percent error. Display all calculated results on a Data Sheet of your own devising, so they can be readily checked.

ANALYSIS AND INTERPRETATION

1. Which do you believe would be more noticeable, a change in the air temperature of one Celsius degree or of one Fahrenheit degree? Give a reason for your answer.
2. If the Celsius temperature T_C were known, the following equation could be used to calculate the Fahrenheit temperature.

$$T_F = (1.8)T_C + 32$$

(a) Solve the above equation for T_C to obtain an equation that could be used to calculate the Celsius temperature when the Fahrenheit temperature is known. Display the equation prominently.
(b) In the above equation, does the number 1.8 have units or is it a dimensionless number?
3. Show that the equation relating T_F and T_C in question 2 can be obtained using Eq. (18.1) and (18.2).
4. Assume the ideal human body temperature is exactly 98.6°F. Express this temperature in Celsius degrees.
5. What would d_2/d_1 be equal to for a room temperature of
(a) 10°C?   (b) 30°C?   (c) 63°F?
6. Liquids other than mercury are often used in thermometers. What are some properties that such liquids should have?

STUDY OF A THERMOCOUPLE

PURPOSE

To study the thermoelectric effect and to plot a calibration chart for a laboratory thermocouple.

APPARATUS

Iron-constantan[1] or copper-constantan thermocouple; millivoltmeter (range depends on metals in thermocouple); beakers; ice; burner; thermometers (0.2°C accuracy); ring stand.

INTRODUCTION

In 1821 T. J. Seebeck (1770–1831) found that, if two different metals such as copper and iron are joined *to form a closed loop* and if one junction is kept at a different temperature from the other, electric current will flow in the closed loop (see Fig. 19.1). This discovery is known as the *thermoelectric effect* or *Seebeck effect*. The electrical potential difference between the two junctions, which can be measured with a sensitive voltmeter, can be used to determine the temperature of one of the junctions if the temperature of the other junction is known. Such an arrangement is called a *thermocouple*.

The usual circuit arrangement (Fig. 19.2) for measur-

---

[1]Constantan is an alloy of approximately equal parts of nickel and copper.

ing temperature with a laboratory thermocouple allows the use of a thermocouple for which neither metal junction is copper. This arrangement allows ordinary copper wire to

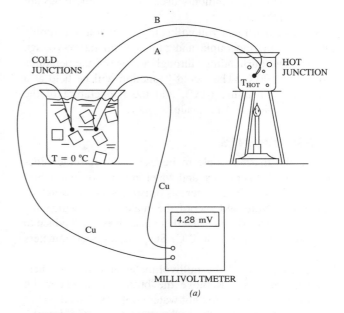

MILLIVOLTMETER

(a)

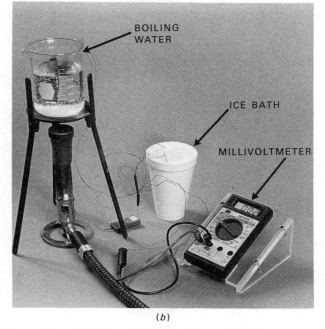

(b)

Figure 19.2. (*a*) Usual circuit arrangement for making temperature measurements with a laboratory thermocouple. Some pairs of typical metals for the thermocouple wires labeled *A* and *B* are listed in Table 19.1. (*b*) Typical laboratory setup for studying a simple thermocouple.

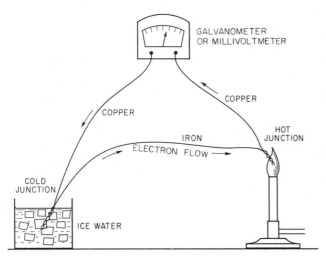

Figure 19.1. The Seebeck effect. When wires of two different metals are joined to form a closed loop, and when the two junctions are kept at different temperatures, there will be an electrical potential difference between the two junctions and an electric current will flow in the closed loop.

Table 19.1   Properties of Four Commonly Used Thermocouples

	Iron-constantan	*Copper-constantan*	*Chromel-alumel*	*Platinum (10% rhodium, 90% platinum)*
		Thermocouple type		
Usual temperature range (°C)	−200–750	−200–400	−200–1200	0–1500
Typical emf (mV) with hot junction at 100°C	5.3	4.3	4.1	0.64

be used between the cold junction and the millivoltmeter. Four of the more commonly used thermocouple types are listed in Table 19.1.

In this experiment you will study the action of a simple laboratory thermocouple and plot a temperature-voltage curve from your readings through a range of temperatures from 0 to 100°C. The "cold" junction will be held at a constant temperature (0°C), and the temperature of the "hot" junction will be made to vary.

MEASUREMENTS

Set up the thermocouple to be used with its cold junction(s) in a beaker of ice and water and its hot junction(s) in a beaker of boiling water, as illustrated in the sketch of Fig. 19.2. Note and record the materials of which the wires are made. Be sure that there is enough cracked ice to hold the cold junction at 0°C. Suspend the thermometers in the beakers.

Read and record the millivoltmeter and the two thermometer readings. Remove the burner from under the boiling water, and as the hot water cools, take readings of the thermometers and the millivoltmeter at 5°C intervals. As the hot water nears room temperature, you may add a little ice to hasten the cooling, but take care that the cooling is uniform. Continue adding chips of ice to the "hot" water until its temperature gradually reaches 0°C, taking

millivoltmeter readings every 5 degrees. If necessary, siphon out some of the water occasionally.

CALCULATIONS

There are no numerical calculations. The results are to be shown in graphical form. Use graph paper, and plot a *neat, carefully labeled* temperature-voltage graph for the thermocouple you used. Think carefully in deciding which variable you will plot vertically, and which horizontally. Note on the graph a description of the thermocouple (i.e., the metals and the number of junctions involved).

ANALYSIS AND INTERPRETATION

Include a brief historical discussion on Seebeck's work.

1. List at least three situations in industry where the use of a thermocouple-type thermometer would be indicated.
2. Describe the steps which would be necessary to calibrate an industrial thermocouple.
3. Approximately what is the sensitivity (in millivolts per degree Celsius per single junction) of your thermocouple for the temperature range studied? Compare this result with accepted table values. How can the sensitivity of thermocouples be increased?
4. How are the thermocouples used in pyrometry?

DATA SHEET

EXPERIMENT 19 ■ STUDY OF A THERMOCOUPLE

Temperature of hot junction (°C)	Meter reading (mV)	Temperature of hot junction (°C)	Meter reading (mV)
Boiling water temp. _____°C			
95		45	
90		40	
85		35	
80		30	
75		25	
70		20	
65		15	
60		10	
55		5	
50		0	

Temperature of cold junction _____0°C (Be sure to hold this
temperature throughout.)

Junction materials _____

EXPERIMENT 20

LINEAR EXPANSION OF METALS

PURPOSE

To study the linear expansion of metals on being heated, and to determine the linear coefficient of thermal expansion for aluminum, copper, and steel.

APPARATUS

Linear expansion apparatus, micrometer screw form[1]; electric buzzer or 6-V (or 12-V) lamp, and wire for connections; source of dc; test rods of aluminum, copper, and steel; steam generator and connecting hoses; thermometer (0.5°C accuracy); laboratory burner, meter stick.

INTRODUCTION

Matter expands on being heated and contracts on being cooled. The amount of expansion and contraction is greatest in gases; it may be quite appreciable in liquids, and is least in solids. However, even in solids, the amount of expansion is usually of such magnitude that it cannot be neglected in the design of machinery and industrial apparatus, particularly if the temperature variation is expected to be considerable.

The phenomenon of thermal expansion is explained by the kinetic-molecular theory of matter. According to the theory, the kinetic energy (and therefore the velocity) of the molecules of a substance increases with increasing temperature. As molecules acquire greater energy, they move with greater speed and collide with one another violently, with the result that their average distance apart tends to become greater. The actual volume (or area or length) of the material thus increases with increasing temperature.

In the case of metal rods or wires it is expansion in length (*linear expansion*) which is of importance. In Fig. 20.1 let L_1 be the length of a metal rod or wire at an initial temperature t_1, and L_2 its length at some higher tempera-

<hr>

[1] The roller-and-dial form of the apparatus may be used, if preferred.

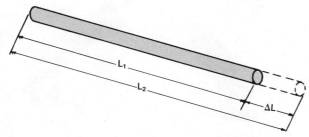

Figure 20.1. Factors involved in linear expansion of a metal rod caused by an increase in temperature. The elongation ΔL is grossly exaggerated.

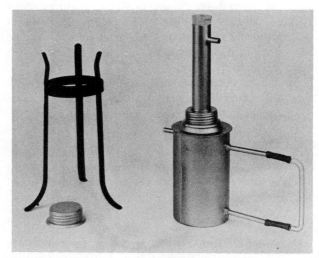

Figure 20.2. Laboratory steam generator (boiler) and support stand. Central Scientific Co.

ture t_2. The *coefficient of linear expansion* is defined as the *change in length per unit of original length per degree change in temperature*. As a formula:

$$\alpha = \frac{L_2 - L_1}{L_1(t_2 - t_1)} \tag{20.1}$$

If the change in length is designated $\Delta L = L_2 - L_1$ and the change in temperature $\Delta t = t_2 - t_1$, then

$$\alpha = \frac{\Delta L}{L_1 \Delta t} \tag{20.2}$$

In the diagram of Fig. 20.1, ΔL is much exaggerated.

The following instructions are for the SI-metric system, but you may use the English system if preferred.

MEASUREMENTS

Fill the boiler (Fig. 20.2) two-thirds full of water and begin heating. Then measure the lengths of the test rods at room temperature. Estimate to the nearest 0.5 mm. This length is denoted by the letter L_1.

Assemble the linear expansion apparatus (Fig. 20.3a) carefully, making necessary electrical connections for buzzer or lamp operation. Carefully zero the micrometer screw. Place one test rod in the (cold) steam jacket, clamp it in place, and take an average of four readings of the micrometer screw (Fig. 20.3b) on the contact point while the rod is still at room temperature. The thermometer should be carefully inserted in its well so that it nearly touches the test rod. It is held in position by a rubber stopper. Read the micrometer at the point where the buzzer just stops sounding (or the point where the lamp goes out). Designate the average of these four readings as x_1. Read and record the jacket temperature. After the x_1 reading is determined, be *very careful* not to jar or disturb the apparatus.

Back off the micrometer screw several millimeters, connect the steam hose to the steam generator, and begin passing steam through the jacket with a rubber-tubing drain to the sink. Be sure steam passes freely through the apparatus. After about 5 min, or when the thermometer reading is constant at or near 100°C, take another series of four micrometer readings on the contact point. Designate the average of these readings as x_2. Read and record the steam jacket temperature.

Disconnect steam lines, disassemble jacket, and cool it thoroughly under the faucet.

Repeat the entire procedure with the other rods.

CALCULATIONS

The net increase in length, or elongation ΔL, of the test rods is given by the difference between the two micrometer readings for each test rod.

$$\Delta L = x_2 - x_1 \qquad (20.3)$$

Determine ΔL from Eq. (20.3). Substitute your data in Eq. (20.2), and calculate the linear coefficient of expansion (α) for each of the test rods. Note that the units of α are

$$\frac{cm}{cm \cdot C°} \quad \text{or} \quad \frac{in.}{in. \cdot F°}$$

depending on the system of measurement used in gathering the data.

Summarize your results in a neat table, and compare your answers with accepted values (from tables) for each of the metals tested. Calculate the percent error of each of your determinations.

ANALYSIS AND INTERPRETATION

Discuss the significant sources of error in this experiment. Why is it necessary to measure ΔL so accurately (with a micrometer screw), whereas the original length L_1 can be measured with a meter stick or yardstick?

List and discuss at least three industrial applications where linear expansion must be taken into account.

Answer the following:

1. Since the units of α are

$$\frac{in.}{in. \cdot F°} \quad \text{or} \quad \frac{cm}{cm \cdot C°}$$

it can be readily seen that these units actually reduce to

$$\frac{\cancel{in.}}{\cancel{in.} \, F°} \quad \text{or} \quad \frac{\cancel{cm}}{\cancel{cm} \, C°}$$

If you knew the value of α per degree Celsius for a metal, how could you calculate the corresponding value per degree Fahrenheit?

2. The suspension span of the Golden Gate Bridge is 4200 ft long. Find the total expansion of the bridge (inches) as the temperature changes from 40 to 85°F. Use the value of α for mild steel given in the Appendix.

3. A structural steel bar 1 m long and 1 cm square is heated to 150°C and its ends are then rigidly clamped. What is the force (newtons) on the clamps developed by the bar as it cools to 20°C? (*Hint:* Review Young's modulus, Experiment 14).

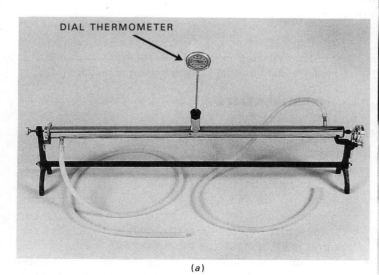

(a)

Figure 20.3. (a) Linear expansion apparatus, micrometer screw form, with dial thermometer. Buzzer (or lamp) circuit not shown. (b) Close-up of micrometer screw dial, which records changes in length.

(b)

Name _____ Date _____ Section _____

Course _____ Instructor _____

DATA SHEET

EXPERIMENT 20 ■ LINEAR EXPANSION OF METALS

Test rod	Original length, L_1 (cm)	Micrometer reading, cold, x_1 (mm) (ave. of 4 readings)	Micrometer reading, hot, x_2 (mm) (ave. of 4 readings)	Cold jacket temperature, t_1 (°C)	Hot jacket temperature, t_2 (°C)
1. _____ (material)					
2. _____ (material)					
3. _____ (material)					

NOTES, CALCULATIONS, OR SKETCHES

EXPERIMENT 21

CHARLES' LAW OF GASES

PURPOSE

To study the relationship between volume and temperature for a given mass of gas confined at constant pressure, and to estimate absolute zero on the Celsius temperature scale.

APPARATUS

Charles' law apparatus; boiling-water bath; ice-water bath (as near 0°C as possible); Celsius thermometers (0.2° precision); thin metal wire.

INTRODUCTION

Charles' law states that the volume of a given mass of an "ideal" gas[1] is directly proportional to its absolute tem-

---

[1] Air, like any other ordinary gas, is not an "ideal gas," but for the temperature range of this experiment, its behavior is very close to being "ideal."

perature T if the pressure on the gas is held constant. Mathematically,

$$V = kT \qquad (21.1)$$

where k is a constant and T is *absolute* temperature, on either the Kelvin or the Rankine scale. An alternate expression for Charles' law is

$$\frac{V_1}{V_2} = \frac{T_1}{T_2} \qquad (21.2)$$

Equation (21.1) is a linear equation. When volume V is plotted against temperature T, a straight line of slope k results. By extrapolating the line determined from known data to a volume of zero, one can determine the temperature associated with zero volume (see Fig. 21.1). Repeated experiments under the most careful conditions give this temperature as $-273.16°C$, or $-459.7°F$. These temperatures are, respectively, *zero degrees* Kelvin and *zero degrees* Rankine. In reality, if one actually tried to per-

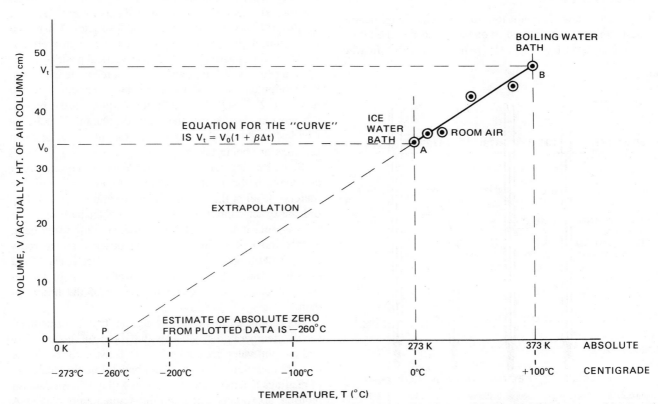

Figure 21.1. A plot of data from a hypothetical Charles' law experiment, showing the *AB* portion of the straight-line graph fitted to a set of (presumed actual) data obtained between $t = 0°C$ and $t = 100°C$; and the *AP* straight line extrapolation to determine the temperature at which the volume V theoretically reduces to zero. Theoretically, the extrapolated dashed line should intersect the temperature axis at $0K$ or $-273°C$.

Charles' Law of Gases **109**

form such an experiment, all gases would liquefy or even solidify before absolute zero was reached. Despite that fact, however, the mathematical extension is a very useful tool in aiding us to understand the laws of gases.

A third mathematical expression for Charles' law is

$$V_t = V_0(1 + \beta \Delta t) \tag{21.3}$$

where V_t is the volume at any Celsius temperature (above absolute zero), β is the coefficient of volume expansion; V_0 is the volume at 0°C; and Δt is the Celsius temperature change above (+) or below (−) 0°C. The value of β has been found to be the same for *all gases*—1/273 per Celsius degree.

The Celsius and Kelvin temperature scales are similar in that a 1° interval on both scales represents the same actual change of temperature. The same can be said of the Fahrenheit and Rankine scales. The Kelvin and Rankine scales are "absolute" scales, since their zero points represent the lowest possible temperature theoretically obtainable, in other words, "absolute zero." All molecules are at their lowest energy state at absolute zero. The relationships among these scales are summarized here and are illustrated in Fig. 21.2.

$$1\ C° = 1.8\ F° \tag{21.4}$$
$$K = °C + 273.16 \tag{21.5}$$
$$°R = °F + 459.7 \tag{21.6}$$

This experiment is to be performed in the SI-metric system, with thermometer readings in Celsius degrees.

MEASUREMENTS

The Charles' law apparatus to be used in this experiment is a thin capillary tube closed at one end and mounted on a centimeter scale (see Fig. 21.3). It contains dry air (the

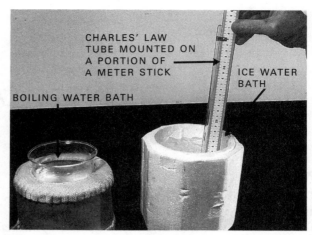

Figure 21.3. The Charles' law tube being lowered into the ice-water bath. Ice must be generously used to insure that the temperature remains at 0°C (273 K). The boiling-water bath is shown at left.

gas whose volume changes will be observed) confined by a "plug" of mercury that is free to move (with negligible friction) as the volume of air changes with changing temperature. Thus, the pressure remains constant as volume and temperature change.

Should the plug of mercury become separated at any time, use the thin wire to get the parts combined again. Do not proceed with the experiment if the mercury becomes separated. If you cannot recombine the mercury, ask your instructor for assistance or for another apparatus. To avoid mercury separation in the tube, be very careful not to jar the apparatus as it is used during the experiment. *Mercury is poison!* If any is spilled notify the instructor. Wash your hands thoroughly if you come in contact with mercury.

Allow the apparatus to stand in a vertical position in the room air for about 5 min while you are readying the rest of the equipment. Record the room air temperature to the nearest 0.2°C, and then the scale reading (to the nearest 0.5 mm) at each end of the trapped column of air in the Charles' law tube. Be sure to keep the tube vertical when readings are being taken.

Place the apparatus in the container of ice and water and leave it there for 2 or 3 min. Read and record the mercury level at the top of the air column. Read and record the exact temperature of the ice-water bath.

Now, place the tube slowly into the boiling water bath. Gently move it up and down and observe the mercury plug. Wait until everything has been at a steady state for 2 or 3 min, and then read the scale at the top of the air column. Record the exact temperature of the hot-water bath.

Repeat these steps with water at two intermediate, steady, temperatures between the room air temperature and the hot-water bath temperature. Be sure to record the exact temperatures of these water baths. Finally, take a set of readings in a tap-water bath. You will now have data for plotting six points on a volume-temperature curve. A blank-form data sheet is purposely not provided. Very carefully, plan your own data sheet.

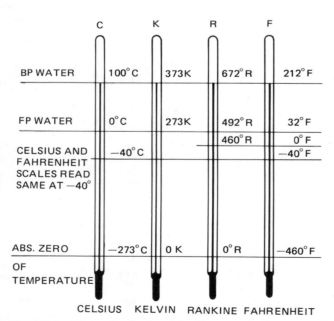

Figure 21.2. Diagram showing the relationships among the Celsius, Kelvin, Rankine, and Fahrenheit temperature scales. The sketch is schematic only, since liquid-in-glass thermometers have only very limited temperature ranges.

CALCULATIONS AND RESULTS

For each set of measurements substract the scale reading at the bottom end of the trapped air column from the scale reading at the top end. This is the height of the air column, and since the glass tube has a uniform cross-sectional area, these heights are actually measures of volume, and they will be so used.

On graph paper plot carefully (review the instructions on plotting graphs in the Introduction) each of the above-determined "volumes" against its corresponding temperature in °C. Choose a temperature scale that runs from −300°C to +150°C. Plot temperatures along the horizontal axis, volumes along the vertical axis. After all six points are plotted, draw the best fitting straight line through the points that you can. (Figure 21.1 is a plot of some hypothetical data, to illustrate the procedure.)

Now extrapolate this line downward and to the left until it intersects the T axis, to determine the temperature at which the volume of trapped air in the tube would theoretically be zero. This is your estimate of absolute zero in degrees Celsius.

ANALYSIS AND INTERPRETATION

1. Why can it be assumed that the pressure is constant during the experiment?
2. Relate this law of gases to Boyle's law (Experiment 17). To Gay-Lussac's law. To the *ideal gas law*.
3. Do some reference reading and write a paragraph or two on the meaning of an "ideal" or "perfect" gas.
4. Can you suggest a couple of ways that the accuracy of this experiment could be improved?

NOTES, CALCULATIONS, OR SKETCHES

EXPERIMENT 22

FUEL VALUE OF NATURAL GAS

PURPOSE

To determine the fuel value or heat of combustion of the laboratory gas supply, using a continuous-flow calorimeter.

APPARATUS

Continuous-flow calorimeter; thermometers (0.2°F accuracy); Fisher-type burner; gas meter with dial readings accurate to 0.01 ft³; rubber tubing; laboratory stand; balance and weights.

INTRODUCTION

Industry uses many fuels; some very common and inexpensive, and others quite rare and costly. Fuel oil, coal, kerosene, gasoline, and natural gas are examples of relatively low-cost fuels. Liquid hydrogen and hydrazine are examples of less common and quite expensive fuels. The first five mentioned are the basis of industrial production throughout the country, while the second group is representative of fuels being used in rocket motors to furnish thrust for missiles and space vehicles.

Some fuels are convenient and safe to use; others are used under very difficult conditions and at relatively great hazard to personnel. Some lend themselves to automatic control of the heating process and leave little ash or residue; others require considerable attention from an operating engineer and leave a bulky residue requiring frequent shutdowns for cleaning.

Although many factors enter into a decision as to which fuel will be used for a specific purpose, one of the most important is the *fuel value* of the fuel in question. Fuel value is defined as the number of heat units which a unit quantity of the fuel will produce. The term *heat of combustion* is also used. In the engineering system of units, fuel value is expressed either in British thermal units (*Btu*) *per pound* (solid and liquid fuels) or in British thermal units (*Btu*) *per cubic foot* (gaseous fuels). Some energy companies express heat units in *therms* (1 *therm* = 10^5 Btu), and state fuel values in therms per hundred cubic feet (therms/100 ft³). In the metric system, fuel value is expressed in calories per gram or kilocalories per kilogram or in kilocalories per cubic meter. The English (engineering) system of units is used in the following instructions. However, the SI-metric system may be used if desired.

In order to measure the heating value of a fuel, an arrangement must be provided whereby all the heat given off by the fuel can be absorbed by some substance. Water is used as the substance for a variety of reasons, chief among which are:

1. It is standard for all heat measurement.
2. It is plentiful, cheap, safe, and readily lends itself to flow measurements.
3. It has high specific heat capacity.
4. It is a very commonly used medium for heat transfer and heat exchange.

This transfer of heat from the products of combustion is carried out in a *heat exchanger* called a *continuous-flow calorimeter*. A *continuous-flow calorimeter* for gaseous fuels is diagrammed in Fig. 22.1. The arrangement is such

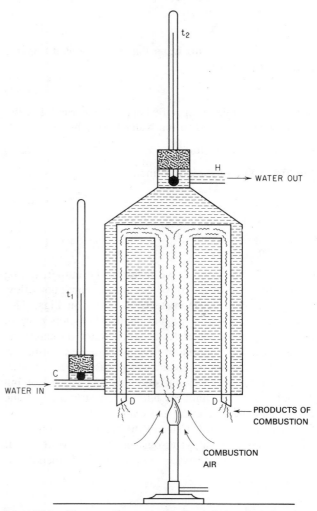

Figure 22.1. Diagram of a continuous-flow calorimeter. It is a heat exchanger of the fire-tube type.

that as much as possible of the heat given off by the burner is absorbed by the water as it flows slowly and continuously through the calorimeter. The tubes *D* force the hot gases of combustion to heat a large surface area of the copper calorimeter, and thus *heat exchange* between the flame and the water is improved. This heat exchanger is known as a *fire-tube* type, to distinguish it from the *water-tube* type in which water circulates in tubes surrounded by the firebox.

Water flows in at *C* and out at *H*, the rate of water flow being carefully controlled in order to obtain a small but constant difference (10 to 25°F) between water-in and water-out temperatures. Since pressure variations in the laboratory water supply are apt to occur, it is recommended that an overflow reservoir (or weir) be used to maintain a constant pressure ''head.'' The test is run for a considerable time during which all the water which flows through the calorimeter is collected, and the gas consumed is carefully metered. From the quantity of fuel used and the change in temperature of the measured weight of water through the calorimeter, the fuel value can be determined. Let

w = weight of water collected, lb
t_1 = water-in temperature, °F
t_2 = water-out temperature, °F
V = ft^3 gas burned

The heat given to the water may be calculated from the basic heat equation

$$H = wc(t_2 - t_1)^1 \qquad (22.1)$$

Remembering that the specific heat *c* of water is 1 Btu/lb · F°, the heat delivered to the water by the burner is

$$H = w(t_2 - t_1) \text{ (in Btu)} \qquad (22.2)$$

The fuel value of the gas is then

$$\text{Fuel value} = \frac{H}{V} \text{ in } \frac{\text{Btu}}{\text{ft}^3} \qquad (22.3)$$

MEASUREMENTS

Connect the gas through the meter to the burner, noting proper direction of gas flow. Set up the continuous-flow calorimeter with thermometers, as shown in Figs. 22.1 and 22.2. The stoppers around the thermometers must fit tightly to prevent water leaks. Turn water on carefully, and allow a *slow* continuous flow through the calorimeter. Adjust the overflow reservoir (weir) to assure steady flow. When the water is flowing smoothly and gas is coming through the meter, light the burner, adjust to a low, blue flame, and place it under the calorimeter. Adjust water flow and flame until a steady difference of about 10°F between water-in and water-out temperatures is obtained. Some care may be necessary to prevent a ''flameout'' of

---
[1] If SI-metric units are used, Eq. (22.1) becomes

$$H = mc(t_2 - t_1) \qquad (22.1')$$

where *m* is the *mass* of water collected, in kg, and *t*'s are in °C.

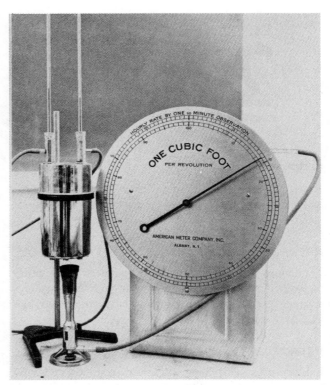

Figure 22.2. Gas meter, burner, and continuous-flow calorimeter setup for fuel value determinations.

the burner. Best results are usually obtained when the burner is about ½ to ¾ in. below the bottom of the calorimeter. Watch the burner constantly during the test to make sure it has not flamed out from the stack effect of the fire tubes.

When a steady state (water flow and temperature readings) has been reached, note and record the gas meter and thermometer readings and simultaneously start catching the water off the calorimeter. Catch a large amount (a half-gallon or more), and record the gas meter reading as you stop catching the water. Weigh the water accurately. Take three more trials, adjusting flame and water flow for temperature rises of about 15, 20, and 25°F.

CALCULATIONS

Using Eqs. (22.2) and (22.3), calculate the fuel value for each trial and get the mean of the four results. Determine from the manufacturer the efficiency of the burner (usually not more than 80 percent), and correct your mean value for burner inefficiency.

From the local gas company, obtain the rated fuel value of the gas and compare your result (corrected for burner losses) with the rated value. Calculate the percent error of your final result.

ANALYSIS AND INTERPRETATION

1. What are the other sources of error (in addition to burner losses) in performing the experiment?
2. Why do piston engines for aircraft require high-grade

fuels such as aviation high-test gasoline, while aircraft jet engines can burn relatively low-grade fuels like kerosene?

3. List four advantages of natural (or manufactured) gas as a fuel, even though it may have a relatively high cost factor in some regions.

4. Heat obtained from fuels is a result of combustion. Just what is the nature of combustion? What does a supercharger (as used on some internal combustion engines) do for combustion?

5. What is "fuel knock," and how can it be controlled in auto engines?

6. Besides the fuel (liquid hydrogen, etc.) what other vital component must rockets and satellite boosters carry with them?

7. If you were the industrial engineer of a large new plant which contemplated using billions of British thermal units per month in a central heating and power installation, what factors would you take into account before deciding on what fuel to use? (Coal, fuel oil, natural gas.) Arrange your list in a table of advantages and disadvantages, which could help you and your employers to make a final decision.

8. What products of combustion contribute significantly to air pollution and environmental change generally? What products are relatively harmless?

9. Coal is the fossil fuel in greatest abundance in the United States. What are some methods of mining, transporting, and processing coal that will make its use more economical and less environmentally destructive?

NOTES, CALCULATIONS, OR SKETCHES

Name _____ Date _____ Section _____

Course _____ Instructor _____

DATA SHEET

EXPERIMENT 22 ■ FUEL VALUE OF NATURAL GAS

Trial	Initial gas-meter reading (ft^3)	Final gas-meter reading (ft^3)	Water-in temp., t_1 $(°F)$	Water-out temp., t_2 $(°F)$	Weight of water w (lb)
1					
2					
3					
4					

NOTES, CALCULATIONS, OR SKETCHES

EXPERIMENT

23

HEAT MEASUREMENT—SPECIFIC HEAT CAPACITY

PURPOSE

To study the law of heat exchange, and to determine the specific heat capacity of sample liquids and solids.

APPARATUS

Calorimeter and stirrer; Celsius thermometer (0.2° markings); burner and stand; beaker; centigram balance; metal test samples; test liquids such as methanol or isopropyl alcohol. *Notes:* (1) A styrofoam cup with paper top and wood stirring stick makes an acceptable calorimeter, if standard laboratory calorimeters are unavailable. (2) The metal test samples should be quite large, to maximize heat transfer from the water in the calorimeter cup.

INTRODUCTION

For many years heat was conceived of as a fluid, called *caloric,* that flowed into and out of bodies or substances, with accompanying temperature increases or decreases. With this concept of the *nature of heat,* the idea of *quantity of heat* was a natural consequence. According to the present-day *kinetic-molecular* explanation of heat energy, however, heat is defined as *energy in transit,* which flows in response to temperature difference. Heat energy flows from a substance or an object at a high temperature to one at a lower temperature. Heat will not flow unless there is a temperature difference. Heat is a manifestation of the *internal energy* of a substance or body.

To say that a substance *contains* a known amount of heat is inconsistent with the kinetic-molecular theory of heat. However, in order to deal with the practical problems of engineering and industry, some quantification of heat energy flow is needed, and the units of heat quantity proposed centuries ago are still in use today. The quantity of heat contained in a substance cannot be measured by any direct reading instrument in the way that a thermometer measures temperature or a gage measures pressure. Heat can be measured only by its effect on matter.

The most readily measured *effect* of heat is temperature change. When heat energy is supplied to various substances, it is observed that the temperature change is different for each different substance. Some substances undergo a marked temperature increase when only a small amount of heat energy is added, while others require much heating before appreciable temperature change occurs.

Water is the standard for heat measurement. In the *mks* system the *kilocalorie* (kcal) is the unit of heat quantity, and it is defined as that quantity of heat required to raise the temperature of *one kilogram* of pure water one Celsius degree. The engineering system unit of heat in the United States is the *British thermal unit* (Btu), and one Btu is defined as that quantity of heat required to raise the temperature of *one pound* of pure water one Fahrenheit degree.

Recalling that 1C° = 1.8 F°, and that the mass equivalent of 1 lb is 0.4536 kg, it can be readily calculated that

$$1 \text{ Btu} = \frac{0.4536}{1.8} = 0.252 \text{ kcal}$$

Since heat is a form of energy, there is at present a definite trend to express heat quantities in standard energy units, namely *joules* (J). (See Experiment 27, Mechanical Equivalent of Heat.)

Heat Capacity   In general, the number of kcal required to raise the temperature of a given mass of a substance by 1 C° is called the *heat capacity* of that substance. From the definition of the kcal, the heat capacity of 1 kg of water is 1 kcal/C°. Water has the greatest heat absorbing capacity of any common substance, and this is one reason it is used as the standard for heat measurements. The heat capacity of water is 1.00 Btu/F° in the English system of units, since 1 Btu of heat will raise the temperature of 1 lb of water 1 F°.

Specific Heat   The heat capacity per unit mass of a substance is the *specific heat* (sp ht) of that substance. Specific heat therefore tells how many kilocalories of heat are required to raise the temperature of 1 kg of a substance 1 C°. (It also tells how many British thermal units are required to raise 1 lb of a substance 1 F°.) Specific heats are expressed in kcal/kg·C° and in Btu/lb·F°.

This definition of specific heat leads to the fundamental heat equation

$$H = mc(t_2 - t_1) = mc\Delta t \qquad (23.1)$$

where

H = heat (in kcal)
m = *mass* (in kg)
c = sp ht
Δt = temperature change (in C°)

For English-system units

$$H = wc(t_2 - t_1) = wc\Delta t \qquad (23.1')$$

where H is in Btu

$w = weight$ in lb
Δt = temperature change (F°)
c = sp ht (as before)

The mks system will be used in this experiment. Express all masses in kilograms (kg).

The *law of heat exchange* says that when substances at different temperatures are mixed, heat flows from the hotter to the colder substances, and that the heat lost by the hotter substances equals the heat gained by the colder substances.

Since water is the standard of heat capacity, a substance whose specific heat is to be determined is ordinarily mixed with or immersed in water. The mixing process must take place in a container, and the heat capacity of the container is also involved. Losses from the container must be minimized. The container for heat capacity experiments is called a *calorimeter* (see Fig. 23.1).

Let

m_h = mass of hotter substance
c_h = sp ht of hotter substance
t_h = temperature of hotter substance
m_c = mass of colder substance
c_c = sp ht of colder substance
t_c = temperature of colder substance
m_{cal} = mass of calorimeter cup
c_{cal} = sp ht of calorimeter material
t_f = final temperature of mixture and calorimeter

For substances mixed in a calorimeter, the *law of heat exchange* may be written

$$m_h c_h (t_h - t_f) = m_c c_c (t_f - t_c) + m_{cal}\, c_{cal} (t_f - t_c) \qquad (23.2)$$

or, in words,

$$\begin{matrix} \text{Heat lost by} \\ \text{hot substance} \end{matrix} = \begin{matrix} \text{heat gained by} \\ \text{cold substance} \end{matrix} + \begin{matrix} \text{heat gained by} \\ \text{calorimeter and stirrer} \end{matrix}$$

MEASUREMENTS

Part 1. Specific Heat of a Liquid

First weigh the inner calorimeter cup (plus stirrer) accurately to 0.01 g, and then fill it about half full of one of the test liquids. Weigh it again to determine the exact mass of the liquid added. Place the inner cup in the calorimeter, fit the lid, insert thermometer, and record the exact temperature, to 0.2°C.

Then pour in enough hot water (note its exact temperature) to fill the cup about three-fourths full. Stir gently and read the final temperature of the mixture. Then weigh the inner cup again to determine the mass of hot water added.

Figure 23.1. Laboratory calorimeter—outer cup, inner cup, stirrer and lid. *(Central Scientific Co.)*

Record the specific heat of the calorimeter cup material (usually stamped on the cup).

Take another trial using different amounts of the test liquid and hot water.

Repeat, making two similar determinations with a different test liquid.

Part 2. Specific Heat of a Metallic Solid

Use cold water from the tap in the calorimeter cup for this determination. Be sure to have enough water to cover the metal test specimens. Let stand for a few minutes and then determine the water temperature to 0.2°C. The calorimeter cup is assumed to be at this same temperature. Also, use metal specimens that are quite large so they will have appreciable heat absorbing capacity. A test specimen of small mass will give poor results.

Heat the aluminum test specimen in boiling water for at least 5 min. The assumption is, then, that its temperature is the same as that of the boiling water. (Measure the boiling water temperature to 0.2°C.) Transfer the test specimen quickly to the cold water in the calorimeter cup, and put the lid on. Agitate gently and note the final temperature of the water in the cup. Be sure to note *all temperatures and masses* necessary for the calculation of the specific heat of the aluminum specimen.

Take another trial with the aluminum specimen, using a different amount of water in the calorimeter.

Repeat with two trials for the steel specimen and two for the copper specimen (or other metals as provided).

CALCULATIONS

Using Eq. (23.2), calculate (for each trial) the specific heat of the test liquids and of the solid test materials. Average the results of the trials for each substance. Look up the accepted values in tables, and compute the percentage error of your results. Display your results in a neat table.

ANALYSIS AND INTERPRETATION

1. Analyze the possible sources of error in this experiment. What steps could have been taken to minimize the effect of heat gains from and losses to the room?

2. Explain how a calorimeter minimizes heat gains and losses from

a. Conduction

b. Convection

c. Radiation

3. List and describe briefly three industrial processes where heat exchange is of vital importance.

4. Name and describe briefly three typical heat exchangers used in industry.

5. Why is water the medium used in so many heat exchange processes? List as many reasons as you can think of.

NOTES, CALCULATIONS, OR SKETCHES

Name _____ Date _____ Section _____

Course _____ Instructor _____

DATA SHEET

EXPERIMENT 23 ■ HEAT MEASUREMENT—SPECIFIC HEAT CAPACITY

Part 1. Specific Heat of a Liquid

Test liquid	Trial	Mass of calorimeter cup (with stirrer), empty (kg)	Mass of calorimeter cup and stirrer, with test liquid (kg)	Mass of calorimeter cup, final (kg)	t_c (°C)	t_h (°C)	t_f (°C)
	1						
	2						
	1						
	2						

Specific heat of calorimeter cup material _____

Part 2. Specific Heat of a Metallic Solid

Test specimen	Trial	Mass of calorimeter cup (with stirrer), empty (kg)	Mass of calorimeter cup with stirrer and tap water (kg)	Mass of metal specimen (kg)	t_c (°C)	t_h (°C)	t_f (°C)
Aluminum	1						
	2						
Steel	1						
	2						
Copper	1						
	2						

NOTES, CALCULATIONS, OR SKETCHES

EXPERIMENT 24

CHANGE OF STATE—LATENT HEAT

PURPOSE

To study change of state processes, and to determine the latent heat of fusion of ice and the latent heat of vaporization of water.

APPARATUS

Calorimeter and stirrer; steam generator with water trap; burner; thermometers (0.2° accuracy); ice; triple-beam balance and weights; laboratory stand; rubber tubing. The experiment can be performed in either SI-metric units (Celsius) or English units (Fahrenheit), as your instructor prefers.

INTRODUCTION

Heat energy is required to convert a solid to a liquid, or a liquid to a gas. Conversely, heat energy is given off as gases condense to the liquid state and as liquids freeze or solidify. These heat-energy transformations occur at the solid-liquid and liquid-vapor boundaries *without a change in temperature*. Since the heat involved produces no *sensible* effect (i.e., temperature change) in the substance, we refer to this energy as *latent* (meaning "hidden") *heat*.

It is the purpose of this experiment to study change of state processes with water as the substance and to determine experimentally the *latent heat of vaporization L_v* and the *latent heat of fusion L_f* of water.

The latent heat of fusion of a substance is the amount of heat required to change unit mass of the substance (already at the melting point) from the solid state to the liquid state, with no change in temperature. The temperature at which this change of state takes place is called the *melting point*.

The latent heat of vaporization of a substance is the amount of heat required to change unit mass of the substance from the liquid state to the vapor state, without a change of temperature. The temperature at which this change of state process most readily occurs is called the *boiling point*.

These change of state processes for water (under a pressure of 1 atm) will be studied, using a calorimeter and the law of heat exchange or the *method of mixtures*.

When two substances at different initial temperatures are mixed, the quantity of heat lost by the hotter body is equal to that gained by the colder body, and an intermediate or equilibrium temperature is eventually reached. The process must take place in such a way that no heat is gained from or lost to the surroundings. The calorimeter will minimize heat gains and losses, but an additional pre-

caution will be observed as follows: In all the tests, the range of temperature change should be *equally below and above room temperature,* so that heat gained from the room while the calorimeter contents are below room temperature will be balanced by heat lost to the room while the contents are above room temperature.

In determining L_f, a few cubes of ice are dropped into a calorimeter in which a known amount of warm water at a known temperature has been placed. As heat exchange takes place, the ice melts. The final equilibrium temperature (when all the ice has melted) is recorded, and the actual mass of ice used is determined by weighing the calorimeter and its contents again. The ice used should be removed from the freezer some time before the experiment begins, so it will be at its melting point (0°C or 32°F) by the time it is to be used.

In a similar manner L_v is determined by passing live steam through a known amount of cool water in a calorimeter. The steam condenses in the water, and a final equilibrium temperature is reached as heat exchange takes place. The actual mass of steam taking part in the heat-exchange process is determined from a re-weighing of the calorimeter and its contents. Working equations may be derived as follows.

Part I. Latent Heat of Fusion—Equations

Let

m_i = mass of ice
m_w = mass of warm water
m_c = mass of calorimeter and stirrer
t_i = temperature of ice (assumed 0°C or 32°F)
t_w = initial temperature of warm water and calorimeter
t_f = final (equilibrium) temperature in the calorimeter
c_w = sp ht of water = 1.00 kcal/kg · C° or 1.00 Btu/lb · F°
c_c = sp ht of calorimeter material
L_f = latent heat of fusion (water)
W.E.$_{therm}$ = water equivalent of the thermometer

(The *water equivalent* of the thermometer is that mass of water which would have the same thermal capacity as the portion of the thermometer submerged in the water. A standard laboratory thermometer immersed for a length of 3 or 4 cm will have a water equivalent of about 0.5 g.)

The law of heat exchange may then be expressed by the equation (remembering that $t_i = 0$°C)

$$m_i L_f + m_i c_w(t_f - t_i) = m_w c_w(t_w - t_f) + m_c c_c(t_w - t_f)$$
$$+ \text{W.E.}_{\cdot\text{therm}} c_w(t_w - t_f) \qquad (24.1)$$

from which

$$L_f = \qquad\qquad\qquad\qquad\qquad\qquad (24.2)$$
$$\frac{(m_w c_w + m_c c_c + \text{W.E.}_{\cdot\text{therm}} c_w)(t_w - t_f) - m_i c_w(t_f - t_i)}{m_i}$$

Careful analysis will reveal that the left-hand side of Eq. (24.1) sums up the heat gained by the cold substances (ice and melted ice), and the right-hand side sums up the heat lost by the hot substances (warm water, calorimeter, and the submerged portion of the thermometer). This is the law of heat exchange.

Part 2. Latent Heat of Vaporization—Equations

Let

m_w = mass of cool water in calorimeter

m_s = mass of steam added

m_c = mass of calorimeter and stirrer

t_w = initial temperature of cool water and calorimeter

t_s = temperature of steam (assumed 100°C or 212°F)

t_f = final (equilibrium) temperature in calorimeter

L_v = latent heat of vaporization (water)

c_w = sp ht of water = 1.00 kcal/kg · C° or 1.00 Btu/lb · F°

c_c = sp ht of calorimeter material

W.E.$_{\cdot\text{therm}}$ = water equivalent of the thermometer

The law of heat exchange results in

$$m_s L_v + m_s c_w(t_s - t_f) = m_w c_w(t_f - t_w) + m_c c_c(t_f - t_w)$$
$$+ (\text{W.E.}_{\cdot\text{therm}} c_w)(t_f - t_w) \qquad (24.3)$$

from which

$$L_v = \qquad\qquad\qquad\qquad\qquad\qquad (24.4)$$
$$\frac{(m_w c_w + m_c c_c + \text{W.E.}_{\cdot\text{therm}} c_w)(t_f - t_w) - m_s c_w(t_s - t_f)}{m_s}$$

MEASUREMENTS

Part I. Latent Heat of Fusion (L_f) of Water

Determine the mass of the calorimeter cup and stirrer accurately,[1] and note the material of which they are made. Pour in enough warm water to half fill the cup. The temperature of the water should be about 10°C above room temperature. Weigh accurately again, and then place cup in the outer calorimeter jacket. Cover with lid, and after stirring, record the temperature of the water to the nearest 0.2°C.

Select a half-dozen cubes of ice about the size of a walnut. *Dry each carefully* (why?) with a piece of paper towel and drop it in the water. Add ice slowly until the temperature in the cup is about 10°C below room temperature, stirring all the while.

---
[1] Or *weigh* them, if English units are being used.

Figure 24.1. Steam generator with delivery hose. A water trap (not shown) should be placed in the steam delivery tube.

Record the final (equilibrium) temperature when all the ice has melted. Weigh the calorimeter and its contents (less thermometer) again.

Repeat the entire process for two more trials, using different amounts of water at different initial temperatures, and different amounts of ice. Take every precaution to minimize heat gains to and losses from the calorimeter contents.

Part 2. Latent Heat of Vaporization of Water

Fill the calorimeter cup about three-fourths full of water about 10°C below room temperature. (Cool it with ice if necessary.) Determine masses and temperature as before.

Adjust the steam generator (Fig. 24.1) for slow but steady flow, and make sure a water trap is in the steam delivery line to trap the condensed steam, thus assuring that "dry steam" will be delivered to the calorimeter. (Why?) Measure the steam temperature at the end of the delivery tube.

Insert the steam tube under the water surface in the calorimeter (lid should be on), and allow steam to pass into the water *slowly* (why?) and condense until the temperature of the water has risen to about 10°C above room temperature. Remove the steam line, stir, read final (equilibrium) temperature, and weigh calorimeter cup and contents again (less thermometer).

Repeat for two more trials, using different initial values of water mass and temperature, and a final temperature as far above room temperature as the initial water temperature is below it. (If English units are used, substitute "weight" for "mass" and °F for °C in the directions.)

CALCULATIONS

1. From the data of Part 1, using Eq. (24.2), compute the latent heat of fusion of water for all trials, and average the results. Express the answer in both metric units (kcal/kg) and English units (Btu/lb). Use dimensional analysis in converting from one system to the other.

2. From the data of Part 2, using Eq. (24.4), compute the latent heat of vaporization of water for all trials, and average the results, again expressing the answer in both systems of units.

3. Summarize your results in a table, and compare with accepted values by computing the percent error of your results.

ANALYSIS AND INTERPRETATION

Discuss the sources of error in the experiment.

Describe in considerable detail at least three change-of-state processes important in industry.

Answer the following:

1. If the chunks of ice were not dry, why would the experimental error be increased?

2. What is the purpose of the water trap in the steam line? Would omitting it tend to make the result for L_v too high or too low?

3. What would be the effect of a pressure other than atmospheric on the values of L_v and L_f?

4. Explain carefully the terms *sensible heat* and *latent heat*.

NOTES, CALCULATIONS, OR SKETCHES

DATA SHEET

EXPERIMENT 24 ■ CHANGE OF STATE—LATENT HEAT

Part 1. Latent Heat of Fusion of Ice, L_f

Trial	Mass of calorimeter cup and stirrer, m_c (kg)	Mass of warm water, m_w (kg)	Initial temp., t_w (°C)	Final temp. mixture, t_f (°C)	Mass of ice added, m_i (kg)
1					
2					
3					

Temperature of ice t_i is assumed 0°C or 32°F

Part 2. Latent Heat of Vaporization of Water, L_v

Trial	Mass of calorimeter cup and stirrer m_c, (kg)	Mass of cool water m_w, (kg)	Initial temp., t_w (°C)	Final temp. mixture, t_f (°C)	Temp. steam, t_s (°C)	Mass of steam added, m_s (kg)
1						
2						
3						

If English units are being used, change *mass* to *weight* (lb) and °C to °F.

Specific heat of calorimeter material, c_c = _____

Room temp = _____ °C

Change of State—Latent Heat **129**

EXPERIMENT 25

HEAT TRANSFER—THERMAL CONDUCTIVITY OF COPPER

PURPOSE

To study heat transfer by conduction, and to determine the thermal conductivity of copper.

APPARATUS

Heat conductivity apparatus (Searle's);[1] thermometers (0.2°F accuracy); steam generator; burner; time clock; balance and weights; catch basin; rubber tubing. (The experiment may be performed using Celsius temperatures and mks units if desired.)

INTRODUCTION

Of the three heat transfer processes—*conduction, convection,* and *radiation*—this experiment deals only with conduction. The manner in which heat is transferred through solids will be studied, and the heat conductivity of copper will be determined.

The kinetic-molecular theory assumes that molecules, even in solids, are in more or less continuous random motion. In solids the actual motion is probably quasi-vibratory, the molecule not actually moving from one location in the solid to another location, but just vibrating or oscillating, constrained to or trapped in its own "domain," surrounded by other molecules. The theory further assumes that heat energy is associated with molecular kinetic energy, or with the speed of the molecules.

The kinetic-molecular explanation of heat transfer by conduction is that as heat energy is absorbed by molecules at the "hot side" of a solid substance, these molecules vibrate violently, setting their neighbor molecules in motion by collision. These neighbors, in turn, pass the energy on to their neighbors, and thus heat energy is passed through the solid.

The amount of heat H which will flow through a substance by conduction depends on five factors, namely:

1. The cross-section area A of the material through which the heat flows
2. The temperature difference $(t_h - t_c = \Delta t)$ between the hot side and the cold side
3. The time T of heat flow
4. The thickness L of the material
5. A constant of the particular substance, its *thermal conductivity, k.*

---

[1] A different design of a thermal conductivity apparatus is available from Pasco Scientific Company.

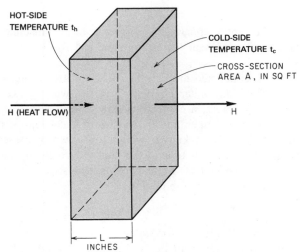

Figure 25.1. Schematic diagram of the relationships involved in heat transfer by conduction.

The instructions for this experiment are in the English system of units.

Figure 25.1 illustrates the factors involved. Theoretical analysis, combined with careful experiments, verify the following basic equation for heat conduction:

$$H = \frac{kAT(t_h - t_c)}{L} = \frac{kAT\Delta t}{L} \qquad (25.1)$$

where H is the heat (Btu) conducted through a substance of cross-section area A (ft^2), in time T (h), when the difference in temperature of the two sides Δt is $(t_h - t_c)$ degrees Fahrenheit. L is the thickness of the material (in.).

Solving for k gives

$$k = \frac{HL}{AT\Delta t} \qquad (25.2)$$

Dimensional analysis reveals that the units of k are [from Eq. (25.2)]

$$\frac{\text{Btu/h}}{(\text{ft}^2)(\text{F°/in.})}$$

To determine k for a test sample, L, A, T, t_h and t_c must be measured, and a means of determining H, the heat flow in time T, must be devised.

The apparatus diagrammed in Fig. 25.2 is designed so that all the factors of Eq. (25.2) may be measured. The test specimen is a copper rod AB, one end of which is placed in a steam jacket, and the other end in a cold water

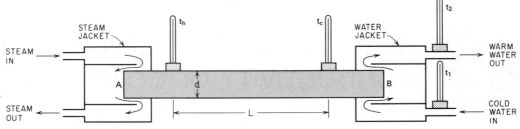

Figure 25.2. Diagram of one form of heat conductivity apparatus.

jacket. The entire apparatus is surrounded by an insulated box to minimize heat losses from the test rod (see Fig. 25.3).

Heat is conducted through the rod from A to B and is given off to the water circulating in the water jacket, which is, in effect, a continuous-flow calorimeter. Cold water enters at temperature t_1, and warm water leaves at temperature t_2. If a known weight of water w passes through the jacket in time T the heat picked up by the water from the end of the test rod in that same time is given by the basic heat equation

$$H = wc(t_2 - t_1) \qquad (25.3)$$

where $c = 1$ Btu/lb $\cdot$ F° for water.

The value of H from Eq. (25.3) is then used in Eq. (25.2), along with the other measured variables, to determine k.

The values of t_h and t_c can be read from thermometers after a steady-state heat flow has been established. L can be measured directly and A can be computed from the measured diameter d of the test rod.

MEASUREMENTS

Set up the apparatus as shown in Fig. 25.4. A few drops of mercury should be placed in the t_h and t_c thermometer wells to assure a good thermal contact. CAUTION! *Mercury is poison*. If any is spilled notify your instructor immediately. Pass steam steadily through the steam jacket and adjust the flow of water through the water jacket to a small stream. A "constant head" overflow device (weir) is recommended, as shown in the illustration. Loosen thermometer t_2 momentarily to eliminate air pockets in the water jacket, then insert the stopper securely again so there are no leaks.

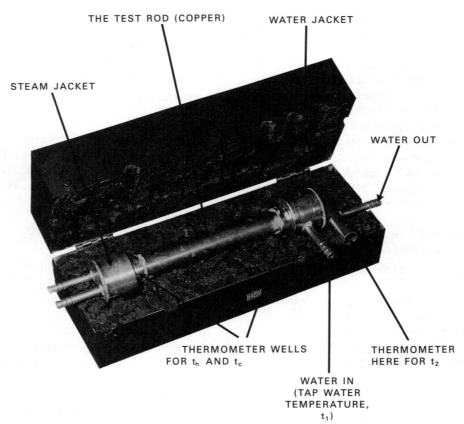

Figure 25.3. Apparatus for determining the heat conductivity coefficient k of copper. Sometimes known as Searle's heat conductivity apparatus. *(Central Scientific Co.)*

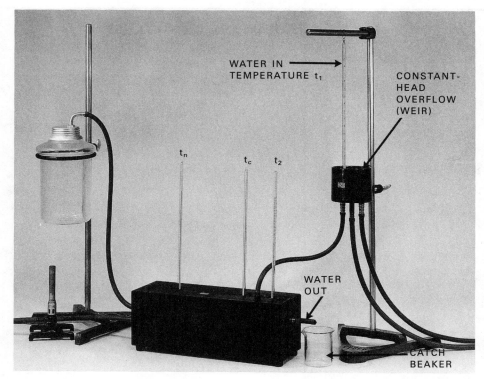

Figure 25.4. Laboratory setup for determining the heat conductivity coefficient of copper. Note the *constant head* overflow device (weir) at right. *(Central Scientific Co.)*

After a steady-state heat flow (as indicated by a constant difference between t_h and t_c) has been attained, adjust the flow of water to give a difference between t_2 and t_1 of about 10°F. Continue observing the readings of all four thermometers until a steady-state condition has been reached and maintained for about 5 min.

When this condition has been attained, read and record $t_h, t_c,$ $t_1,$ and $t_2,$ and catch all the warm water flowing out of the water jacket for a measured time interval $T,$ say, 10 min. The longer the duration of the trial, the more reliable the results will be. Watch the thermometers closely throughout the test run. If marked fluctuations occur, start the trial over again. Determine the weight of the water collected and record the time T and the readings of the four thermometers.

Take three additional sets of data under different conditions, varying the flow of water to obtain values of $(t_2 - t_1)$ of about 15, 20, and 30°F.

Measure the distance L between the t_h and t_c thermometers, and the diameter d of the test rod—both to an accuracy of 0.01 in. Record all data on the Data Sheet.

CALCULATIONS

Determine the value of $H,$ the heat transferred to the water during each trial, from Eq. (25.3). Substitute these values in Eq. (25.2), and calculate k for all four trails. Take care to express all values in the proper units, as defined following Eq. (25.1). Average your results and compare with the correct value of k for copper, as listed in tables. Calculate the percent error of your answer.

ANALYSIS AND INTERPRETATION

1. Analyze the sources of error in this experiment.
2. Summarize what you have learned about heat transfer in solids.
3. Discuss the relative advantages of copper and aluminum for heat transfer purposes (e.g., in cooking vessels, heating and cooling coils for air conditioning units, automobile radiators, and heat exchangers for industrial processes). Take into account all significant factors.
4. k is a heat *conductivity* factor (often called k-factor) for *homogeneous* materials. Based on some reference reading, define and discuss ''R-factors'' and ''U-factors'' (for nonhomogeneous materials). How are they used in the building materials industry and in the refrigeration and air conditioning industry?
5. A cast iron heat exchanger in a furnace has an effective surface area of 12 ft^2 and is 0.28 in. thick. If the temperature of the iron on the fire side is 350°F and that on the air side is 338°F, find the heat transfer rate in Btu per hour.

DATA SHEET

EXPERIMENT 25 ■ HEAT TRANSFER—THERMAL CONDUCTIVITY OF COPPER

Trial	Temp. water on, t_1 (°F)	Temp. water off, t_2 (°F)	Temp. hot side, t_h (°F)	Temp. cold side, t_c (°F)	Weight w of water collected (lb)	Time T of run (h)
1						
2						
3						
4						

Length of test rod, L _____ in.

Diameter of rod, d _____ ft

NOTES, CALCULATIONS, OR SKETCHES

EXPERIMENT 26

RADIANT ENERGY VS. SURFACE TEMPERATURE OF A RADIATING BODY

PURPOSE

To study the relationship between the absolute (Kelvin) temperature of the tungsten filament in an incandescent bulb and the power radiated by the filament.

APPARATUS

Autotransformer to provide an adjustable ac voltage; ac ammeter; ac voltmeter; 100-W incandescent bulb with socket; wires for connections; ohmmeter, or digital multimeter.

INTRODUCTION

Heat energy can be transferred from one location to another by *conduction, convection,* or *radiation.* Of these three heat transfer mechanisms, it is only by radiation that energy can be transferred through a vacuum or through space. For example, it is by radiation that energy from the surface of the sun travels at the speed of light through 93 million miles of space (vacuum) to the earth.

Energy, in the form of electromagnetic waves, is radiated from the surface of any object whose temperature is above absolute zero (0 K). The rate at which energy is radiated from the surface of an object depends on the area A and on the absolute temperature T of the surface in a manner predicted by the equation

$$P = \sigma \epsilon A T^4 \tag{26.1}$$

where

P = radiated power in watts
σ = Stefan–Boltzmann constant
    = 5.67×10^{-8} W/m$^2 \cdot$ K^4
ϵ = emissivity, a dimensionless number between
    0 and 1 that is equal to the
    ratio of radiant emittance
    (or absorptance) of a surface to
    that of the surface of an
    ideal blackbody. (ϵ for a blackbody = 1.00)
A = surface area of radiating object in m^2
T = surface temperature of radiating
    object in kelvins.

In this experiment the radiating object will be the filament of a 100-W incandescent bulb. Since it would be difficult to make a direct measurement of the temperature of the filament, an indirect method will be used.

The resistance R (in ohms) of any cylindrical conductor depends on its length L (in cm) and its cross-sectional area A (in cm^2) as predicted by the equation

$$R = \rho \frac{L}{A} \tag{26.2}$$

where ρ is the *electrical resistivity* in ohm-centimeters ($\Omega \cdot$ cm) of the conductor material.[1] Since resistivity is a temperature-dependent property of the material, the resistance R will also depend on temperature. For the filament in this experiment, the resistance R_0 and resistivity ρ_0 at room temperature are related to the resistance R and the resistivity ρ at an elevated temperature by the equation

$$R_0/\rho_0 = R/\rho \tag{26.3}$$

For resistivities between 50×10^{-6} $\Omega \cdot$ cm and 80×10^{-6} $\Omega \cdot$ cm, the filament temperature in kelvins can be calculated using the equation

$$T = \frac{\rho + 1.20 \times 10^{-5}}{3.434 \times 10^{-8}} \tag{26.4}$$

The operating voltage V and the current I (amperes) for the filament can be used to calculate the power P (watts) radiated by the filament and the resistance R (ohms) of the filament using the equations

$$P = VI \text{ (volts} \times \text{amperes)} \tag{26.5}$$
$$R = V/I \tag{26.6}$$

Equations (26.3) and (26.6) can be used to determine the resistivity of the hot filament, and Eq. (26.4) can then be used to calculate the absolute temperature of the filament.

A plot of values from Eq. (26.5) for the power radiated by the hot filament, versus the fourth power of the calculated temperature (Eq. 26.4) can be used to confirm the power-temperature relationship given by Eq. (26.1).

MEASUREMENTS

Measure and record the room-temperature resistance R_0 of the filament in the 100-W bulb (see Fig. 26.1). Then connect the autotransformer, ammeter, voltmeter, and light bulb to construct the circuit shown in Fig. 26.2 (see also the illustration of the apparatus in Fig. 26.3).[2]

Obtain data for the operating *voltage* of the bulb and its associated filament *current,* for operating voltages from 40 V ac to 110 V ac. Construct your own data table with

---

[1] Equations (26.2), (26.3), and (26.4) are given here, without derivation.
[2] Take *care in working with 120* V ac!

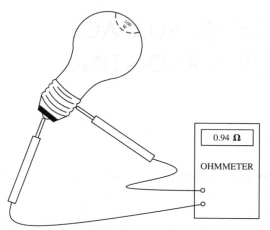

Figure 26.1. Illustration of procedure for measuring the room-temperature resistance of the filament in the 100-W bulb.

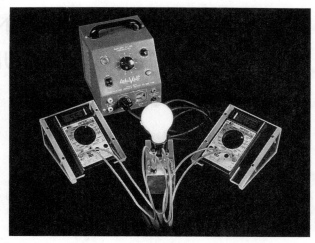

Figure 26.3. Arrangement of apparatus for the circuit shown in Fig. 26.2. The socket for the 100-W bulb is mounted in a small metal box that also contains banana plug jacks to allow convenient and safe connections to the meters used to measure voltage and current. *(Photo by MSOE Academic Media Services.)*

columns 1 and 2 for these measured values for the voltage and current. Provide additional columns for *calculated* values of the power P (Col. 3), the resistance R (Col. 4), the resistivity ρ (Col. 5), the filament temperature T (Col. 6), and the fourth power of the filament temperature T^4 (Col. 7).

CALCULATIONS

Use Eqs. (26.5) and (26.6) to calculate the radiated power and the resistance of the hot filament for each operating voltage. Record the results of these calculations in your data table.

The room-temperature resistivity of the filament[3] in a standard tungsten lamp is given as $5.5 \times 10^{-6}\ \Omega \cdot \text{cm}$. Eq. (26.3) can now be written

$$\rho = \left[\frac{5.5 \times 10^{-6}\ \Omega \cdot \text{cm}}{R_0}\right]R \qquad (26.7)$$

Use Eq. (26.7) to calculate the resistivity of the tungsten filament for each of the operating voltages, and then use Eq. (26.4) to calculate the filament temperature at each operating voltage. Record the calculated values for ρ, T, and T^4 in the proper columns of your data table.

---

[3] From *The Practicing Scientist's Handbook* by Alfred J. Moses, Van Nostrand Reinhold Company, 1978, p. 677.

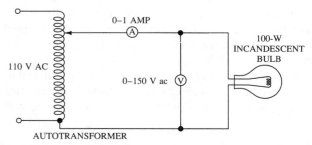

Figure 26.2. Wiring diagram for the circuit used to obtain voltage-current data for the 100-W bulb.

ANALYSIS AND INTERPRETATION

1. On a sheet of graph paper plot the radiated power P along the vertical axis and the fourth power of the filament temperature T^4 along the horizontal axis. Draw a "smoothed curve" through the plotted points. The locus of plotted points should approximate a straight line. What does this tell you about the relationship of radiated power P to T^4?

2. By what factor must the temperature of a surface be increased to cause the radiated power to increase by a factor of 16?

3. If Eq. (26.3) is an acceptably correct statement about the mathematical relationship between resistance and resistivity at different filament temperatures, what must be true about the influence of thermal expansion on the filament? (*Hint:* Review the units of ρ, the resistivity.)

4. The resistance of the filament in a typical incandescent bulb operating at its rated power is approximately 12 times as great as the resistance of the cold (room temperature) filament. Use your data to calculate the ratio of the resistance of the filament in your bulb at an operating voltage of 110 V ac to its resistance at room temperature. Is the ratio approximately equal to 12?

5. If the filament resistance at room temperature T_0 is R_0, then its resistance R at temperature T is

$$R = R_0\ [1 + \alpha(T - T_0)]$$

where α, the *mean temperature coefficient of resistance* of tungsten, is equal to $0.0053\ \text{K}^{-1}$. Calculate an experimental value for α using your data at room temperature and at a filament temperature associated with an operating voltage at or near 100 V. Compute the percent difference between your value and $0.0053\ \text{K}^{-1}$.

6. Use Eq. (26.1) to calculate the surface area A of the filament in your bulb. Calculate A for two different filament temperatures. (Assume the emissivity ϵ of tungsten is 0.23.)

27 MECHANICAL EQUIVALENT OF HEAT

PURPOSE

To study the relationship between heat and mechanical work and to determine Joule's constant, the mechanical equivalent of heat.

APPARATUS

Mechanical equivalent of heat apparatus (Cenco); mercury-in-glass thermometer, accurate to 0.2°C; 5-kg slotted mass; centigram balance.

INTRODUCTION

Mechanical work and heat are two different forms of energy. Heat can be converted into mechanical energy by such devices as the steam turbine and the internal combustion engine. Mechanical energy is converted into heat where work is done against friction, or when gases are compressed.

Energy transformations also occur around us every day, due to natural causes. A good example of a heat-to-mechanical energy conversion is found in water power. Heat from the sun evaporates water from the oceans. Winds (caused by uneven heating of the earth) carry the moisture aloft where it may be deposited as rain or snow at high elevations. The kinetic energy available in unit mass of the water as it flows down a stream or river is a part of the mechanical energy equivalent of the heat energy which was expended on the water by the sun.

Benjamin Thompson (Count Rumford) was perhaps the first to suspect the relationship between heat and mechanical work. As a supervisor of a Bavarian ordnance works (about 1790), he studied the heat produced by friction as large drilling bits were used in the boring of brass cannon. He came to the conclusion that the heat produced by the boring was the result of mechanical work being done.

It was almost 50 years later, however, before the numerical value of the *mechanical equivalent of heat* was determined. James Prescott Joule, in 1843, arranged a set of paddles which could be rotated in water by a set of falling weights. The vessel was insulated to prevent heat gains from or losses to the surroundings. The amount of heat created by the churning paddles was calculated from the temperature rise of the known mass of water, and the amount of mechanical work done was equivalent to the loss in potential energy of the falling weights. Over hundreds of trials Joule found that the ratio of mechanical work done *(W)* to the heat produced *(H)* was always the same.

This ratio is known as the mechanical equivalent of heat, or *Joule's constant,* and is expressed by the equation

$$J = \frac{W}{H} \tag{27.1}$$

Joule's constant is one of the constants of nature, like *g,* the acceleration of gravity. Its numerical value depends, of course, on the system of units used.

$J = 778$ ft · lb/Btu   or   4186 joules/kcal

where

1 joule (J) = 1 newton-meter (N · m)

In this experiment mks units will be used for all measurements and computations.

METHOD

(The procedures given here apply to the Cenco hand-crank, rotating-drum apparatus, as shown in Fig. 27.1. Another apparatus, commonly known as the Cenco-Puluj design, is still in use in some college laboratories. If the Puluj equipment is used, the instructions provided by the supplier should be followed.)

Figure 27.1. Mechanical equivalent of heat apparatus, showing copper drum calorimeter, thermometer, friction band, hanging weight, and crank. *(Central Scientific Co.)*

The apparatus is essentially a copper drum acting as a *calorimeter,* around which are wound several turns of a heavy, braided nylon cord. Each winding of the cord is in direct frictional contact with the copper drum as the latter is rotated by a hand crank. The copper drum is attached to a sturdy bearing in a base plate of non-heat-conducting material which is clamped to the corner of a laboratory table (Fig. 27.1). Provision is made for a thermometer to be inserted in the drum, along the drum's rotational axis. A leak-proof opening is provided for this purpose. The thermometer rotates with the drum and continuously records the temperature of the water in the drum.

One end of the nylon friction cord is hooked into a 5-kg mass used as a hanging weight. The other end of the cord is attached either to a counterweight or to a coil spring which will provide tension on the cord.

You do work as you crank the copper calorimeter (drum) round and round. The friction between the surface of the copper drum and the several turns of the nylon cord is the *force* being overcome by the effort on the hand crank. This frictional force is exactly that required to lift and support the hanging 5-kg mass above the floor. The product of this force and the distance the drum surface moves during an entire trial run is the work done in that trial.

The mechanical work done (in joules) during a trial run is given by

$$W_{(joules)} = \pi d N M g \tag{27.2}$$

where

d = diameter of drum, meters
M = mass of load = 5 kg
g = 9.81 m/s^2
N = revolutions turned

The heat (in kcal) produced by the work done against friction during the trial run is determined from calorimetric data, and can be expressed as follows:

$$H = m_w c_w (t_f - t_i) + m_c c_c (t_f - t_i) \tag{27.3}$$

where

m_w = mass of water in the copper drum, kg
m_c = mass of copper calorimeter drum, kg
c_w = sp ht of water = 1.00 kcal/kg · C°
c_c = sp ht of copper = 0.092 kcal/kg · C°
t_f = final temperature of the water and drum, °C
t_i = initial temperature, °C

The very small amount of heat absorbed by the thermometer and by the nylon band is neglected in Eq. (27.3).

From Eqs. (27.2) and (27.3) the mechanical equivalent of heat J may be calculated from Eq. (27.1).

MEASUREMENTS

Follow the steps listed below very carefully. You are to devise your own Data Sheet for this experiment.

1. Determine the mass of the copper calorimeter, with-

out thermometer, but with locking nut, after cleaning and polishing it.

2. Determine the diameter of the copper drum (calorimeter).

3. Fill the calorimeter with 50 to 60 g of water which is at least 5 C° below room temperature, *and reweigh.* Insert the thermometer through the rubber seal, and tighten locking nut.

4. Attach the calorimeter to the insulated base plate, and clamp the entire apparatus to the corner of the laboratory table.

5. Wind the nylon friction band around the cup for 4 to 6 turns, which will fill the space available on the cup. (Five turns is usually sufficient.) Secure one end of the band with the counterweight just behind the revolution counter on the apparatus, leaving a short loop. Move the counterweight to a position just under the calorimeter. Secure the other end of the friction band to the 5-kg mass on the floor. Adjust the number of turns of the friction band on the calorimeter until turning the crank handle steadily causes the 5-kg mass to be lifted off the floor and kept at a constant height above the floor. (Have instructor check this adjustment!)

6. The water temperature in the drum should be about 3 C° below room temperature. If it is too cold, turn the crank and warm it up a bit. Read and record the initial temperature t_i, and then commence the first trial run.

7. Turn crank rather rapidly but carefully for exactly 300 revolutions. This should cause a temperature rise of about 6 C°, enabling the final temperature to be about 3 C° *above* room temperature. Thus radiation and convection losses are minimized. Be sure that the 5-kg mass is lifted off the floor throughout the entire trial run, by the force of friction.

8. Read the final temperature t_f.

9. Take three more trials, varying the amount of water, the initial temperature, and the number of turns for each trial.

10. Disassemble the apparatus carefully, dry everything, and put it away.

CALCULATIONS

For each trial, calculate the work done from Eq. (27.2) and the heat generated from Eq. (27.3). Then calculate experimental values of J from Eq. (27.1). Obtain the mean of your four results. Calculate the percent error of your mean result from the value given in tables.

ANALYSIS AND INTERPRETATION

1. Analyze carefully the sources of error in the experiment.

2. Discuss at some length the implications of the concept of mechanical equivalent of heat to industry and technology.

3. How does the value of J enter into the determination of the efficiency of a heat engine (steam or internal combustion)?

4. Answer the following:

a. Why is a uniform speed of rotation unnecessary?

b. Explain how mechanical work is accomplished in the experiment when the load is not actually moved through any vertical distance.

5. Although the method used to compensate for conduction and radiation losses and gains gives acceptable results, it is not one of great precision. Why?

6. A 2000-kg auto is traveling on a level road at 100 km/h when the accelerator is released and the brakes are applied. Compute the total heat (kcal) developed in the brake drums during the time the car is brought to a stop. Why is it not necessary to know the time or the distance involved? (Assume the entire stopping force is supplied by friction in the brake drums and lining.)

AIR CONDITIONING AND PSYCHROMETRY

PURPOSE

To study the basic principles of psychrometry and to learn how they are applied in air conditioning.

APPARATUS

Sling psychrometer; Alluard's dew-point apparatus; psychrometric chart; Fahrenheit thermometers (O.2°F accuracy); ethyl ether.

INTRODUCTION

Psychrometric Processes Air is a mixture of gases whose average composition is nitrogen (78 percent), oxygen (21 percent), carbon dioxide (0.04 percent), other gases as traces, and water vapor. Water vapor is the most variable of the constituents of air. Its presence varies from only a trace in "dry air" to as much as 0.03 lb of water vapor in 1 lb of air. When air contains as much water vapor as it can hold at a given temperature, it is said to be *saturated*. The higher the air temperature, the greater the amount of water vapor that air can contain before becoming saturated. Any drop in temperature of saturated air will cause some of the water vapor to condense to a liquid. Thus *dew* is formed in the chill of early morning. *Humidity* is the general term used to indicate the presence of significant amounts of water vapor in the air.

Human comfort is significantly affected by the humidity of the air. Air that is too dry often irritates the nasal passages and chaps the skin. Because perspiration evaporates rapidly from the skin into dry air, one feels cooler in dry air than the actual thermometer reading would justify. By the opposite token, humid air, already near saturation, does not absorb perspiration from the skin, and the discomfort of dank, "heavy" air is felt. The surrounding air is judged to be much warmer than it really is, in terms of human comfort. The expression, "It isn't the heat, it's the humidity" is often heard.

Humidity in the air is expressed by two different measures, as discussed in the following paragraphs.

The scientific study of air–water vapor mixtures is called *psychrometry*. The engineering and technical methods of providing air at temperature and humidity levels conducive to human comfort comprise the science of *air conditioning*. The "condition" of the air in a room or space is determined by a number of factors, all of which play a part in determining whether or not the room's occupants judge themselves to be "comfortable." Water vapor in the air is a factor of prime importance.

Five properties of moist air are to be studied in this experiment. Each is related to the others in ways that will be clarified as you perform the experiment.

1. Dry-Bulb Temperature *(DB)* This is the temperature of air as read from an ordinary thermometer.

2. Wet-Bulb Temperature *(WB)* This is the temperature of air as read from a thermometer whose bulb is covered by a wick wet with water. The thermometer bulb is cooled by the evaporation of water from the wick. The amount of cooling is a function of the rate of evaporation, which in turn depends on how nearly saturated the surrounding air is. The difference between the DB temperature and the WB temperature is called the *wet-bulb depression*. At saturation, the wet-bulb *(WB)* thermometer reads the same as the dry-bulb *(DB)* thermometer, since at saturation, no evaporation from the wet wick will occur. *DB* temperature equals *WB* temperature at saturation.

3. Dew-Point Temperature *(DP)* This is the temperature at which water vapor in the air begins to condense to a liquid. At the dew point air is saturated, and $DB = WB = DP$.

4. Specific Humidity *(W)* This is the moisture content of air in grains of water vapor per pound of dry air (1 lb = 7000 grains) at the *DB* temperature stipulated.

5. Relative Humidity *(RH)* This measure of humidity is that read from humidiguides and similar instruments. It is expressed as a percent and is actually described by the ratio

$$\text{Percent } RH = \frac{\text{specific humidity at the time}}{\text{specific humidity at saturation}} \times 100$$

(Recall that air is *saturated* with moisture when it contains all the water vapor it can "hold" at its present temperature.)

All of the properties of moist air are listed in extensive tables in physics and engineering handbooks. For our purposes, however, a chart that graphically portrays the relationships among the above five factors will be used. The *Carrier Psychrometric Chart* for normal temperatures (reduced size) is included as Fig. 28.3. Study the chart. Note the scale of dry-bulb *(DB)* temperatures along the bottom, the scale of wet-bulb *(WB)* temperatures curving upward and to the right, the scale of specific humidity (W) curving upward along the right-hand side of the chart, and the lines of relative humidity (RH) starting at the lower left and curving upward and to the right inside the chart. The wet-bulb scale (at left) is also the scale of dew-point (DP) temperatures, and it is also the *saturation line,* the locus of

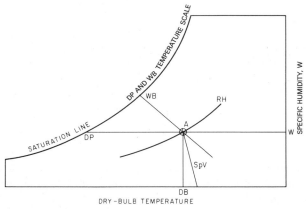

Figure 28.1. Skeleton psychrometric chart showing the relationships among five basic factors in psychrometrics.

all condition points at which the air is saturated with water vapor.

Disregard all other scales portrayed on the chart. They are required for advanced analyses of the conditions of air but will not be referred to in this experiment.

The skeleton chart of Fig. 28.1 will help you in reading and interpreting the psychrometric chart. If any two of the factors defined above are known (such as the DB temperature and the WB temperature) the other three factors can be readily determined. For example, with a known DB and a known WB, determine the intersection of these two lines as A on the chart (Fig. 28.1). A is known as the *condition point* of the air at that time. From A, interpolate between the RH lines sloping upward to the left to determine the relative humidity of the air. Also from A move horizontally to the right and read W, the specific humidity, from that scale. Move left and horizontally from A to the saturation line and read the dew-point *(DP)* temperature.

Dry-bulb temperature, wet-bulb temperature, and dew-point temperature are the most easily *measured* of the five factors, and it is these three that will be determined experimentally.

MEASUREMENTS

In both of the following exercises, use thermometers that are in very close agreement in dry air.

1. With the sling psychrometer (Fig. 28.2a), determine the dry-bulb temperature *(DB)* and the wet-bulb temperature *(WB)* of the air in the room, and record. Then make a similar determination for the outdoor air. Be sure the wick of the wet-bulb thermometer is well soaked with water throughout your determinations. Continue "slinging" until the maximum wet-bulb depression is attained. Take care not to strike anything with the rotating instrument! Take three trials and average them.

2. Fill the cup of the dew-point apparatus (Fig. 28.2b) about half full of ether, and put the cover on it. Polish the outer surface. Then pump the rubber bulb repeatedly to "refrigerate" (chill) the metal cup. Watch the outer cup surface for the *first indication* of lost sheen caused by condensing water vapor on the cup surface from the surrounding air. Immediately read the DB temperature (thermometer in the outside air) and the DP temperature (thermometer in the cold ether). Take three trials and average them.

CALCULATIONS AND RESULTS

Prepare tables which are extensions of the tables on the Data Sheet for both room air and outside air. These tables should be provided with columns added and specific humidity W and relative humidity RH.

Making use of the principles diagrammed in Fig. 28.1 and using the psychrometric chart in Fig. 28.3, determine the factors not measured in the laboratory, and fill in the blank spaces of your two tables.

ANALYSIS AND INTERPRETATION

In the report discuss at considerable length the importance of psychrometry to the science and practice of air conditioning. Summarize what you have learned about the "condition" of air from the experiment.

Based on reference reading, discuss the factors involved in human comfort with respect to the condition of surrounding air.

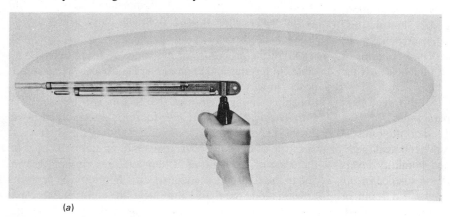

(a)

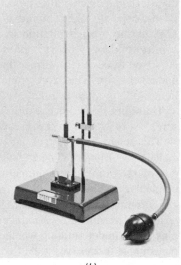

(b)

Figure 28.2. (a) Sling psychrometer—a dry-bulb thermometer and an identical thermometer fitted with a water-soaked wick, both mounted on a frame which can be swung through the air. (b) Alluard's dew-point apparatus.

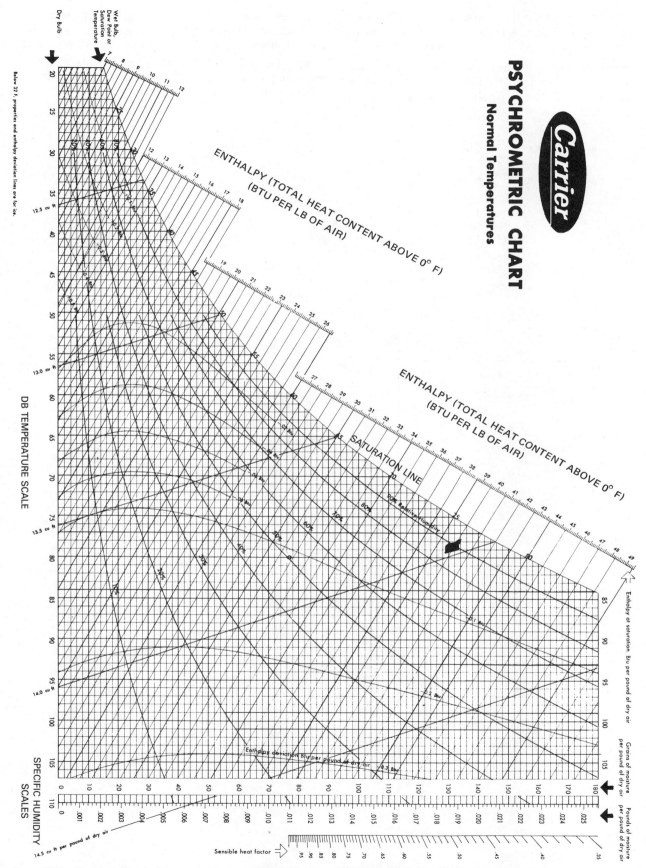

Figure 28.3. Psychrometric chart for normal temperatures, ranging from 20°F DB to 110°F DB. (Carrier Corp.)

Submit the solutions to the following problems as a part of your laboratory report. Use the psychrometric chart as required.

1. A round duct 18 in. in diameter is carrying 2000 cfm (ft³/min) of conditioned air. Find the average velocity of the airstream in the duct in ft/min.

2. Conditioned air is supplied from a room register at 58°F *DB* and 90 percent *RH*. It is discharged into a room whose *DB* temperature remains steady at 80°F. What is the relative humidity in the room? *Hint:* This is a constant moisture process, i.e., *W* is a constant. Use Fig. 28.3 to solve the problem.

DATA SHEET

EXPERIMENT 28 ■ AIR CONDITIONING AND PSYCHROMETRY

Psychrometric Processes

Table I.   Room Air

Method	DB	WB	DP
Sling psychrometer			
Alluard's dew-point apparatus			

Table II.   Outdoor Air

Method	DB	WB	DP
Sling psychrometer			
Alluard's dew-point apparatus			

NOTES, CALCULATIONS, OR SKETCHES

EXPERIMENT 29

WAVE MOTION—FREQUENCY OF A VIBRATING FORK

PURPOSE

To study some principles of wave motion, and to determine the frequency of a vibrating fork by two different methods.

APPARATUS

Vibrograph, complete with tuning fork, pendulum, and sliding plate; stop clock; electrically driven fork or vibrator; braided cord, pulley, and set of weights; meter stick; triple-beam balance, or precision digital balance.

INTRODUCTION

Wave motion may be either *transverse* (as with water waves or vibrating strings) or *longitudinal* (like sound waves in air), or a combination of both. In this experiment transverse wave motions only will be studied. In one case

they will be in the form of a trace made by a stylus attached to a vibrating tuning fork, and in the other, they will be observed as "standing waves" in a string which is set in vibration by an electrically driven fork, In both of these studies the frequency of vibration of the fork will be determined. The frequency of vibration of a vibrating body is defined as the number of complete vibrations per second, vps, or hertz (Hz). A complete vibration is often called a *cycle*. Frequency, then, may be expressed as cycles/s, vib./s (vps), or Hz.

The SI-metric system of units is to be used in this experiment.

Part I. Frequency of a Tuning Fork— Vibrograph Method

The vibrograph (Fig. 29.1) is a device which has both a tuning fork and a pendulum mounted on the same frame.

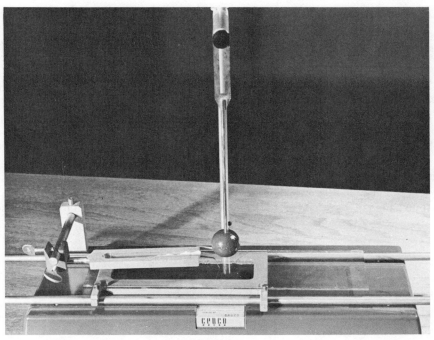

Figure 29.1. Vibrograph with sliding glass plate for recording wave tracings. The tuning fork (left) and the pendulum bob (center) are set in vibration simultaneously. Each carries a stylus that records its wavy motions on a coated glass plate, which is pulled slowly and steadily under the two styli. *(Central Scientific Co.)*

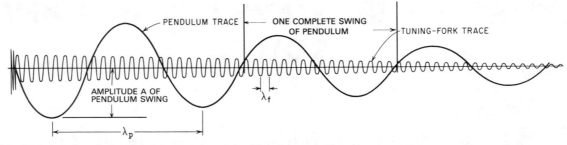

Figure 29.2. Typical tracings obtained on the coated glass plate of the vibrograph. These show transverse wave motions.

Each is fitted with a stylus which traces a record of the swing or vibration on a coated glass plate. The arrangement is such that, when the plate is moved under the two styli, a record like that shown in Fig. 29.2 is obtained. (The plate must first be coated with a thin film of carbon black or Bon Ami powder in alcohol suspension.)

Let N_p be the number of *complete pendulum swings clearly discernible* on the trace. Disregard uncertain "waves" at beginning and end of trace. A *complete* swing begins as the wavy line crosses the center line, and ends when it crosses the center line again going *in the same direction* (see Fig. 29.2) Let N_f be the number of *complete tuning fork swings* contained within the N_p swings of the pendulum. Then N_f/N_p equals the average number of tuning fork vibrations per pendulum vibration.

If, then, the frequency of the pendulum f_p is determined by an actual count with a stop clock, the frequency of the fork f_f may be calculated from the expression

$$f_f = f_p \times \frac{N_f}{N_p} \tag{29.1}$$

Part 2. Frequency of a Fork—Standing-Waves Method

One of the characteristics of wave motion—*frequency*—has been defined above. A second, *amplitude* (A) is the greatest distance from the central position that the wave moves in its cycle (see Fig. 29.2). A third attribute of wave motion is *velocity* **v,** the speed and direction in which the wave is propagated.

"Standing" waves or stationary waves are produced when two waves of the same wavelength, velocity, and amplitude travel in opposite directions through the same medium. A ready means of producing standing waves is by setting up a wave disturbance in a stretched string. A vibrating fork will set up a train of waves at one end of the

string. These waves pass down the stretched string and are reflected back on themselves at the other end. The returning waves react with the oncoming waves to produce standing waves in the string. The situation is illustrated in Fig. 29.3.

The following definitions are of importance:

An *antinode* (or loop) is a point of maximum displacement of the wave.

A *node* is a point of little or no vibration.

One *wavelength* (λ) is the distance from one antinode to the next alternate antinode, or from one node to the next alternate node. In other words, a wavelength is the length of a *complete vibration,* from a point on the wave to the next point on the wave which is in the *same phase of vibration.*

If the length, tension, and mass of the string are such that the returning reflected wave reaches the vibrating fork at exactly the right instant in the fork's cycle, it will be re-reflected with the new wave the fork is then sending down the string. A condition of *resonance* then exists, with greatly increased amplitude.

A stretched string can vibrate in a number of different ways. If it vibrates as a whole, its length equals one-half the wavelength of the vibrations produced. If it vibrates in two segments with a node at each end and one in the middle, the *wavelength of the vibrations produced is equal to the string's length.* In all possible modes of vibration the length of the string is equal to some multiple of half wavelengths of the vibrations produced.

The frequency of vibration which causes standing waves in a cord is given by the expression

$$f = \frac{1}{\lambda} \sqrt{\frac{T}{M}} \tag{29.2}$$

where

f = frequency, vibrations per second (vs) or hertz (Hz)

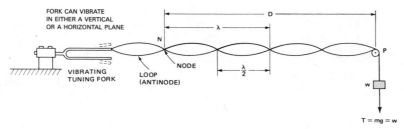

Figure 29.3. Diagram of standing waves in a stretched string which is forced to vibrate at one end.

T = tension in string, newtons
M = mass per unit length of string, kg/m
λ = wavelength of standing waves, m

MEASUREMENTS

Part 1   Set up the vibrograph (Fig. 29.1) so that the styli of both the tuning fork and the pendulum touch the plate lightly. See that the styli are close together and that they vibrate across a common center line. Practice setting the tuning fork and pendulum in vibration simultaneously until you can do it successfully every time.

Coat several glass plates with a thin suspension of Bon Ami powder in alcohol, and allow to dry (or coat with a thin film of lampblack from a burning candle). The result should be an even, thin coating, easily scratched by the styli.

Place one of the coated plates on the sliding platform of the vibrograph, and make final adjustments to the tuning fork and pendulum. Set them both in vibration, and slowly pull the glass plate along under them as they vibrate. A trace like that of Fig. 29.2 should result.

Repeat until you obtain three plates with good, clear traces.

For each plate mark off the number of *complete* pendulum swings clearly discernible on the trace, N_p. Then count the number of *complete* tuning fork swings contained in this distance, N_f. For each of the plates determine the value of the ratio N_f/N_p.

Determine the frequency of the pendulum by timing with a stop clock the interval required for 30 complete swings. Get the average of the three trials and designate this value f_p.

Part 2   Set up the electric tuning fork or vibrator and cord as shown in Figs. 29.3 and 29.4. Let the length of the cord be about 4 m. P is a pulley, and m is a load which puts the cord in tension with $T = mg$.

Operate the vibrator, and adjust the tension until the cord vibrates in two segments. Make minor tension adjust-

ments until the loops formed are of maximum amplitude (a condition of resonance).

Measure D accurately from the pulley to the point of the first node, N, from the vibrator. Count the number of loops between the point N and the pulley. Recalling that the distance from a node to the next node is $\lambda/2$, calculate an average of λ for the standing waves.

Take four more trials, varying the tension to make the string vibrate in three, four, five, and six segments. (Tension must be diminished to produce a greater number of segments.) For each trial make minor tension adjustments by adding or removing small weights to obtain the maximum possible amplitude.

Determine the frequency of the vibrator (if it is not 120 Hz), using an impulse counter and stop clock or a stroboscope with a frequency counter.

Finally, measure accurately a suitable length of the cord (say 4 m when under slight tension) and weigh this amount *carefully* on a precision laboratory balance. Then compute the mass per unit length, M.

Record all data on the Data Sheet.

CALCULATIONS

Part 1   Determine the average value of the ratios N_f/N_p for the three trials, from Table I of the Data Sheet. Calculate the mean frequency of the pendulum f_p from the three trials of Table II. Using Eq. (29.1), compute the frequency f_f of the tuning fork. Obtain the correct value of its frequency from the instructor, and calculate the percent error of your result.

Part 2   Determine the mass per unit length M of the cord in *kilograms per meter*. Using Eq. (29.2), compute the frequency of the vibrating fork for each of the five trials. Average the results, and compare this mean value with the actual frequency of the vibrating fork. Compute your percent error.

Arrange all calculations in a summary table, neatly displayed.

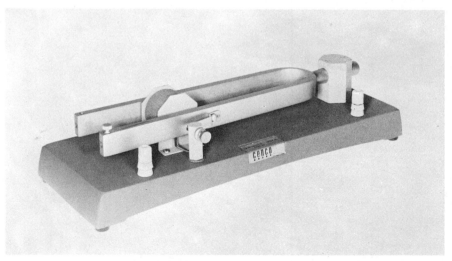

Figure 29.4.   Electrically driven fork used for obtaining standing waves in a stretched string under tension. (*Central Scientific Co.*)

ANALYSIS AND INTERPRETATION

1. Discuss the differences between transverse waves and longitudinal waves. (Only transverse waves were studied in this exercise.)

2. Plot a curve on graph paper from the data of Part 2, using the square roots of the tensions as abscissas and the wavelengths as ordinates. Locate all five points, and connect them with a smooth curve. What law of vibrating strings could you deduce from the shape of this curve?

3. Write a paragraph on vibration in industrial machinery, and discuss some methods of minimizing it. Why must a condition of *resonance* be avoided as machine elements vibrate in proximity to each other? How can resonance be avoided?

4. A steel wire 1 m long with $M = 8.5 \times 10^{-3}$ kg/m vibrates in two segments under a tension of 5.4 N. Find the frequency of the fork causing the vibration. [Careful with units! See Eq. (29.2).]

Name _____ Date _____ Section _____

Course _____ Instructor _____

DATA SHEET

EXPERIMENT 29 ■ WAVE MOTION—FREQUENCY OF A VIBRATING FORK

Part 1. Tuning Fork—The Vibrograph Method

Table I. Data from Traces on Sliding Plates

Trial	No. of complete pendulum swings (N_p)	No. of complete tuning-fork swings (N_f)	Ratios (N_f/N_p)
Plate 1			
Plate 2			
Plate 3			

Table II. Frequency of the Pendulum

Trial	Time for 30 complete swings (seconds)	Average of three trials (f_p = _____ Hz)
1		
2		
3		

Part 2. Frequency from Standing Waves in a String

Trial	Number of segments	Distance (D) meters	Average length of segments ($\lambda/2$)	Average wavelength λ (meters)	Hanging mass (kg)	String tension (N)
1	2					
2	3					
3	4					
4	5					
5	6					

Length of cord weighed   _____ meters

Mass of cord, to nearest 0.00001 kg   _____ kg

Frequency of vibrating fork   _____ Hz

EXPERIMENT 30

SPEED OF SOUND BY RESONANCE METHODS

PURPOSE

To study acoustical resonance, and to measure the speed of sound in air and in metal rods.

APPARATUS

Glass resonance tube about 3 cm in diameter and 1 m long; one-hole rubber stopper to fit tube; laboratory stand and water reservoir with rubber tubing for connection to resonance tube; three tuning forks (suggested frequencies, 384, 440, and 512 Hz); Kundt's tube resonance apparatus with metal rods, cork dust, rosin, and chamois or woolen cloth; meter stick, or steel rule; thermometer.

INTRODUCTION

Obviously, the most direct method of measuring the speed of sound in air would be to station two experimental groups at some distance apart (1 km or more) with visual contact between the groups, and actually determine the time required by sound waves to traverse a measured distance. This can be done with considerable accuracy using present-day electronic timing methods. Air temperature and wind effects must, of course, be taken into account, and appropriate correction factors applied to the measured results.

The same methods could be used to determine the speed of sound in metals or in water, but these methods are understandably not well suited to the space limitations of the standard physics laboratory. This experiment will therefore make use of less direct methods.

The fundamental equation of wave motion is

$$v = f\lambda \qquad (30.1)$$

where

v = velocity (speed) of propagation of the wave
f = number of complete waves per second (frequency)
λ = wavelength

In this experiment the phenomenon of *resonance* will be used in two different investigations—the first, to determine the speed of sound in air, and the second, to determine the speed of sound in metal rods. The meaning of *resonance* and the conditions that bring it about were briefly discussed in Experiment 29. See your physics textbook for detailed explanations.

Note: Sound travels by *longitudinal* waves.

Part I. Speed of Sound in Air

If a vibrating source (such as a tuning fork) is held directly over an air column in a closed tube, compressions and rarefactions will travel down the tube and be reflected (see Fig. 30.1). If the tube length L is adjusted until it is equal to exactly one-fourth the wavelength of the tone from the fork, the returning wave will arrive back at the top of the tube precisely *in phase* with the next vibration of the fork, and a tone of unusually loud volume will be heard. This phenomenon is known as *resonance,* and it occurs when standing waves are set up in a closed tube with a node at the closed end and a loop (or *antinode*) very near the open end (see Fig. 30.1). This situation can occur when the length of the closed tube is any *odd number of quarter wavelengths* of the sound waves being emitted by the fork. Resonance will occur when

$$L = \frac{\lambda}{4}, \frac{3\lambda}{4}, \frac{5\lambda}{4}, \text{ etc.}$$

The apparatus used in the experiment consists of a glass tube about 1.5 m long and 4 cm in inside diameter, fitted with a rubber stopper and a tubing connection to a water reservoir whose level can be changed to make water rise or fall in the resonance tube itself. The apparatus is shown in Fig. 30.2.

Since the frequency of the fork is known (specified by the manufacturer) and λ can be calculated from measurements taken at points of resonance, the velocity of sound in air can be calculated from Eq. (30.1).

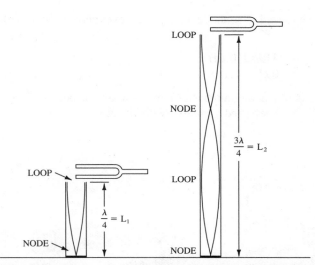

Figure 30.1. Diagrams showing the relationship between wavelength of sound in air and the distance to the first and second points of resonance for a closed tube. The waves are diagrammed, for convenience, as if they were transverse. In fact, sound waves are longitudinal waves.

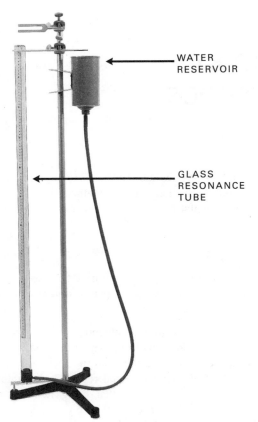

Figure 30.2. Apparatus for determining the velocity of sound in air by resonance methods. *(Central Scientific Co.)*

Part 2. Speed of Sound in Metal Rods

Kundt's method of measuring the velocity of sound in metals depends upon setting up standing waves in the air inside a glass tube. These waves in the air are created by *longitudinal* vibrations in a metal rod which is positioned axially within the glass tube (see Fig. 30.3). Longitudinal vibrations (i.e., lengthwise) along the rod are transferred to the air column in the glass tube by a loosely fitting piston in, but not touching, the tube.

The form of the stationary waves in the air may be deduced from observations of the configurations into which cork dust falls as the air in the tube vibrates in resonance with the natural vibration frequency of the metal rod. The rod itself is stroked longitudinally with a rosin-impregnated chamois or woolen cloth. There may be

a variety of cork dust patterns, but regardless of the particular pattern observed, the distance between corresponding points of adjacent configurations will be *one-half a wavelength of the sound wave in air*. Figure 30.4 shows the relationships involved in a typical pattern.

Derivation of Equations Let f equal the frequency of vibration of the metal rod. This will also be the frequency of vibration of the air in the tube, since the air is forced to vibrate in resonance with the rod. Let

v_a = velocity (speed) of sound in air
v_r = velocity (speed) of sound in the metal rod
λ_a = wavelength of sound in air
λ_r = wavelength of sound in the rod
L = length of the rod

From the fundamental wave equation [Eq. (30.1)],

$$v_r = f\lambda_r$$

and

$$v_a = f\lambda_a$$

Dividing,

$$\frac{v_r}{v_a} = \frac{f\lambda_r}{f\lambda_a}$$

or

$$v_r = v_a\frac{\lambda_r}{\lambda_a} \tag{30.2}$$

If the rod is clamped at its midpoint (thus making a node at the midpoint, and antinodes, or loops, at either end of the rod),

$$\lambda_r = 2L$$

Finally, for the speed of sound in the rod,

$$v_r = v_a\frac{2L}{\lambda_a} \tag{30.3}$$

The speed of sound in air v_a at sea level may be calculated from

$$v_a = 331.7 \pm 0.61t \text{ m/s } (t \text{ in C}°)$$
$$= 1087 \pm 1.1t \text{ ft/s } (t \text{ in F}°) \tag{30.4}$$

where t is the number of degrees above or below 0°C or 32°F. The speed increases with rising temperature.

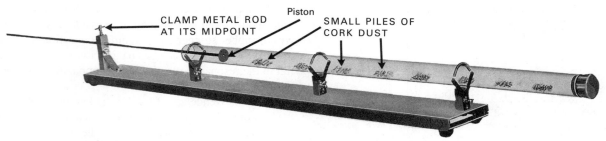

Figure 30.3. Kundt's tube apparatus for measuring the velocity of sound in metal rods. *(Welch Scientific Co.)*

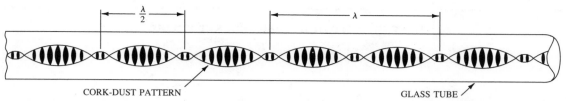

Figure 30.4. Diagram of one typical cork-dust pattern obtained with the Kundt's tube apparatus. The pattern you obtain may not be of exactly this configuration.

This experiment may be done with either metric or English units.

MEASUREMENTS

Part I. Speed of Sound in Air

Raise the water level in the glass tube until it is near the top. Position the tuning fork at the top of the glass tube so that it will barely clear the glass while vibrating. Make sure it will not strike the glass while vibrating. Strike it with a rubber hammer. While it is vibrating, lower the water level slowly until the first resonance point is reached. (The sound will be quite loud at this point, even though the sound of the fork itself may be barely audible.) By approaching the resonance point *carefully* from above and below, determine its position accurately, from the scale on the meter (or yard) stick.

Measure L_1, the distance from the top lip of the glass tube to the first point of resonance.

In like manner determine L_2 and L_3, the distances to the second and third points of resonance, if the tube is long enough. Repeat the entire process with two other forks of different frequencies.

Record on the Data Sheet the frequencies of the forks used and the temperature of the air in the resonance tube.

Part 2. Speed of Sound in Metal Rods

Clamp one of the test rods exactly at its midpoint, in the clamp provided with Kundt's apparatus. One end of the rod is fitted with a metal or felt disk which acts as a loosely-fitting piston in the glass tube, but it must not touch the glass. This end is inserted into one end of the long glass tube (see Fig. 30.3).

The glass tube should be stoppered at the other end and should contain enough cork dust or lycopodium powder to cover very lightly the bottom curved surface of the tube. Its supports should allow it to be moved horizontally to make adjustments which will allow the best possible condition of resonance to be obtained.

Set the rod in vibration by stroking longitudinally the free end, slowly but vigorously, with the rosin-impregnated cloth. A continuous "singing" (but metallic) tone will be heard. Adjust the position of the glass tube by

sliding it longitudinally until the best possible cork dust pattern is produced.

Excluding the half-wave at each end, measure the distance spanned by the maximum number of cork dust patterns that are clearly recognizable as such, and divide by the number of (complete) patterns to get the mean value of $\lambda/2$.

Scatter the cork dust, and repeat the determination a second and a third time, tabulating the results.

Repeat the entire process with several rods of different material.

Measure the temperature of the room air and the lengths of the rods used.

CALCULATIONS

Part 1 From the measurements taken with the resonance tube, calculate values for the speed of sound in air for each of the three tuning forks used. Determine the mean value for the speed of sound in the room. Compare your experimental value (percent error) with the correct value for air at the recorded temperature, calculated from Eq. (30.4).

Part 2 From the measurements taken with the Kundt's tube apparatus, and using Eqs. (30.3) and (30.4), calculate the velocity of sound in the metals used for the test rods. Compare your results (percent error) with accepted values for the metals used from tables.

ANALYSIS AND INTERPRETATION

1. Give a complete analysis of the sources of error in the experiment.
2. Based on collateral reading, write a brief discussion of sonar, and explain how variations in the velocity of sound in ocean water are taken into account.
3. Write a paragraph explaining your interpretation of the phenomenon of resonance. What are the necessary and sufficient conditions for resonance?
4. The famous tenor Caruso is said to have been able to shatter a wineglass by holding a sustained note at a certain pitch. Explain how this might be theoretically possible. Would the same note (pitch) suffice for any wineglass?

Name _____ Date _____ Section _____

Course _____ Instructor _____

DATA SHEET

EXPERIMENT 30 ■ SPEED OF SOUND BY RESONANCE METHODS

Part 1.   Speed of Sound in Air

Trial	Frequency of fork, vib/s (Hz)	Distance to first resonance point, $L_1 = \lambda/4$	Distance to second resonance point, $L_2 = 3\lambda/4$	Distance to third resonance point, $L_3 = 5\lambda/4$
1				
2				
3				

Air temperature _____

Part 2.   Speed of Sound in Metal Rods

Kind of rod (steel, brass, etc.)	Trial	Distance spanned by cork dust pattern	Number of cork dust configurations	Length of a half-wave, $\lambda/2$
_____ Length _____	1			
	2			
	3			
_____ Length _____	1			
	2			
	3			
_____ Length _____	1			
	2			
	3			

Air temperature _____

EXPERIMENT
31

VIBRATING STRINGS—THE SONOMETER

PURPOSE

To study the sonometer, and to verify the mathematical laws of vibrating strings.

APPARATUS

Sonometer with hanging masses; two steel test wires of different diameters; laboratory balance and weights; tuning forks (256, 384, 440, and 512 Hz); rubber hammer.

INTRODUCTION

The vibration frequency of a stretched cord or wire is given by the equation

$$f = \frac{1}{\lambda} \sqrt{\frac{T}{M}} \qquad (31.1)$$

where

f = frequency, Hz
λ = wavelength, m
T = tension, newtons (N)
M = mass per unit length, kg/m

When vibrating in the simplest way (i.e., when the *fundamental* tone is being emitted), the length of the string $L = \lambda/2$. The frequency of the fundamental tone is therefore

$$f = \frac{1}{2L} \sqrt{\frac{T}{M}} \qquad (31.2)$$

Eqs. (31.1) and (31.2) are given here without derivation. See your textbook for their mathematical development.

A *sonometer* (sounding box) will be used to verify Eq. (31.2) and also to study certain other properties of vibrating strings. The sonometer (see Fig. 31.1) consists of a resonating box on which wires may be stretched. Provision is made for an easy determination of the force which stretches the wires. The length of the vibrating portion of a wire may be varied by the use of wooden bridges which can be inserted under the wire at any desired point. A meter scale is mounted on top of the sonometer to facilitate length measurements.

The sonometer is used to tune a wire to the "pitch" of a tuning fork of known frequency. The "tuning" may be effected by changing either the length L or the tension T or both. Wires of different diameters and materials may be used to test the effect of mass per unit length M on the frequency.

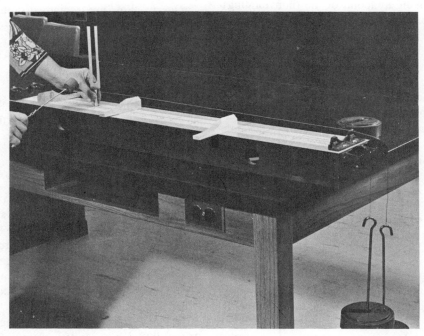

Figure 31.1. Sonometer assembled to compare the frequency of a vibrating string to the known frequency of a tuning fork. The experimenter here is testing for sympathetic vibrations.

There are several ways of determining when the vibrating wire has the same frequency as the test tuning fork:

1. Method of Beats Alternately pluck the string and strike the fork, listening carefully to both. Adjust string until unison is near, at which time "beats" will be heard. Then continue minor adjustment of the string until the "beat frequency" is less than about 1/s.

2. Sympathetic Vibrations When unison is judged to be near "by ear," strike the fork, and place it firmly base down on the sonometer top. See if the strings starts vibrating "in sympathy" with the fork. "Kill" the fork vibration, and listen for the faint continued tone of the wire.

3. Rider Method Place a very small paper *rider* on the wire *at its midpoint,* and then put the base of the vibrating fork down on the sonometer top. If a condition of resonance exists, the paper rider will be thrown violently off the string.

MEASUREMENTS

The mks system will be used for this experiment.

Part I. Verifying the Basic Law of Vibrating Strings—$f = \dfrac{1}{2L}\sqrt{\dfrac{T}{M}}$

Stretch the heavier of the two wires on the sonometer with a load of 1 kg.[1] Adjust the bridges for a wire length L of 30 cm. Use the middle C (256-Hz) fork. Set it in vibration by hitting it with the rubber hammer, and then place it firmly, base down on the sonometer. Pluck the string, and adjusting the tension by changing the hanging mass, gradually bring the string to the same fundamental frequency as the fork. Strike the fork as needed to keep it vibrating.

When you think the tone is about the same, test by the method of beats, and also by the "rider" method. Put a tiny paper rider over the midpoint of the (nonvibrating) string, and place the vibrating tuning fork base down on the sonometer. If the natural frequency of the string is the same as that of the fork, the string will be set in motion and the paper rider will be thrown off. Adjust the string tension until this occurs. Record the total load stretching the wire to determine $T = mg$.

Determine the mass of the wire to the nearest 0.01 g and from its total length determine M, the mass per unit length.

Repeat with the smaller wire ($L = 30$ cm) and a 512-Hz tuning fork. Determine the mass of the smaller wire, its total length, and M, its mass per unit length.

Part 2. Relation of Frequency and Length if Tension and Mass per Unit Length Are Kept Constant

With the *smaller* wire loaded as in the last trial of Part 1, note the length required for a frequency of 512 Hz. Now, leaving the tension load constant, change the length until

the pitch of the string matches that of the 440-Hz fork. Again vary the length until the 384-Hz fork is matched. Finally, match the 256-Hz fork. Check unison each time by all three of the above methods.

Part 3. Relation of Frequency to Tension if Length and Mass Per Unit Length are Kept Constant

With the smaller wire bridged for a length of 30 cm adjust the tension load until the 512-Hz fork is matched.

Now decrease tension until 440 Hz is matched. Then match 384, and finally, 256 by further decreasing the tension. Remember that T (newtons) = mg.

Record all data on the Data Sheet.

CALCULATIONS

Part 1 Calculate the mass per unit length of both wires. Then for both wires substitute the measured values in Eq. (31.2) to check the basic law of vibrating strings. (Take care with units!)

Compute the percent difference of your calculated result from the actual tuning-fork frequency, as specified by the manufacturer.

Part 2 Compute the *reciprocals* of the lengths of the string. Plot a curve using length reciprocals as abscissas and the corresponding frequencies as ordinates. From this curve deduce *and state* a mathematical law for a vibrating string *under constant tension.*

Part 3 From the data in Part 3 calculate the *square roots* of the tension values. Plot a curve using square roots of tension as abscissas and corresponding frequencies as ordinates. From the curve deduce and state a mathematical law for a vibrating string *whose length is constant.*

Query: Why use *reciprocals* of length, and *square roots* of tension?

ANALYSIS AND INTERPRETATION

1. Summarize the laws of vibrating strings that you have discovered in the experiment.

2. Analyze the sources of error in this experiment.

3. Discuss the relationship among the following:

Resonance
Sympathetic vibrations
Beats
Beat frequency

List and discuss briefly applications of these four phenomena to industrial machinery and to structures.

4. *Loudness, pitch,* and *quality* are subjective judgments about sounds. On what physical property does each depend?

5. A wire with a mass a per unit length of 0.005 kg/m is 40 cm long and is vibrating (fundamental) at 256 Hz. What is the tension in newtons?

6. A piano string 1.4 m long, having mass per unit length $M = 0.02$ kg/m, is stretched under a tension of 500 N. What is its fundamental frequency?

---

[1] The suggested loads may be varied as needed to suit the characteristics of the wires being used.

DATA SHEET

EXPERIMENT 31 ■ VIBRATING STRINGS—THE SONOMETER

Part 1. Verifying Eq. (31.2), the Basic Law of Vibrating Strings

Wire	Length of vibrating wire, L (m)	Load on wire (kg)	Frequency of tuning fork (Hz)	Total length of wire (m)	Total mass of wire (kg)
Large steel					
Small steel					

Part 2. Tension and Mass per Unit Length Constant (Small Wire)

Frequency, Hz	512	440	384	256
Length, L (m)				

Part 3. Length and Mass per Unit Length Constant (Small Wire)

Frequency, Hz	512	440	384	256
Load on wire, kg				
Tension, T = mg (N)				

NOTES, CALCULATIONS, OR SKETCHES

PART FIVE ■ LIGHT AND OPTICS

EXPERIMENT 32

PHOTOMETRY AND ILLUMINATION

PURPOSE

To study some principles of illumination, and to measure the light energy output and efficiency of typical incandescent electric lamps.

APPARATUS

Optical bench; bunsen-type laboratory photometer; standard lamp of known power; tungsten filament test lamps of the following wattages: 15, 25, 40, 60, 75, 100, 150; photoelectric foot-candle (lux) meter; black screen (or dark room) for shielding out extraneous light; yardstick or steel measuring tape; large protractor.

INTRODUCTION

The need for effective indoor lighting in schools, libraries, hospitals, offices, factories, and other types of buildings is well known. Also important is the need for adequate outdoor lighting for streets, highways, parking lots, and sports stadiums. Good lighting for these locations results in fewer accidents and may reduce the incidence of crime. In industry, adequate lighting tends to reduce accidents and to increase production.

Homes, too, should have adequate lighting, and many improvements have been made in home illumination since the incandescent lamp was invented by Thomas Edison in 1879. Today's incandescent lamp with its tungsten filament and inert-gas atmosphere is a great improvement over Edison's crude original with its carbon filament and meager light output.

Much of today's commercial and industrial lighting is accomplished by fluorescent lamps or by the newer quartz-halogen, quartz-iodine, tungsten-halogen, sodium arc, and mercury arc lamps, since they generally provide higher efficiencies (more lumens per watt) than the standard tungsten-filament incandescent filament lamps.

In this experiment we shall study some principles of illumination and measure the light-emitting capacity of some incandescent electric lamps. The science of light measurement is called *photometry*.

Certain basic definitions are essential to the study of photometry and illumination. We include a brief summary of these definitions and concepts here. A more extended treatment can be found in your physics textbook.

The Standard Source and Luminous Flux The visible portion of the radiant energy emitted by a lamp is called *luminous flux (F)*. The unit of luminous flux is the *lumen,* but it cannot be accurately defined without first describing the *international standard source* and introducing the idea of a *solid angle* measured in *steradians*.

For many years the standard source was an actual candle manufactured to stipulated specifications, and the unit *source intensity* was the *standard candle*—the intensity of a whale-oil candle burning at the rate of 120 grains per hour. By international agreement in 1948, however, the standard source was redefined in terms of the melting point of platinum. The new *international candle* was defined as one-sixtieth of the luminous intensity of one square centimeter of a blackbody radiator at the temperature of melting platinum (2046 K). The amount of luminous flux emitted from such a source depends on the area of the platinum surface exposed and on the magnitude of the *solid angle* or cone of radiation (see Fig. 32.1).

Imagine a point P on the glowing platinum surface, with a set of lines radiating out in all directions from this point. Now envision a spherical surface of radius r, with its center at point P (Fig. 32.1). The lines radiating out from P cut the surface of the sphere and define a spherical

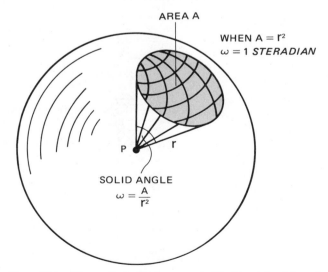

AREA A

WHEN $A = r^2$
$\omega = 1$ *STERADIAN*

SOLID ANGLE
$\omega = \dfrac{A}{r^2}$

Figure 32.1. Diagram illustrating the concept of "solid angle" and the unit of solid angle, the steradian.

surface whose area is A. The solid angle ω which results is defined by the ratio

$$\omega = \frac{A}{r^2} \qquad (32.1)$$

When the area $A = r^2$, the solid angle at P is said to be *one steradian*. The total surface area of a sphere is $4\pi r^2$, and consequently the total solid angle around a point is 4π steradians.

We now define the *lumen* as the *luminous flux (F) emanating from one-sixtieth square centimeter of the standard source* (melting platinum) *within a solid angle of one steradian*.

The *lumen* is a unit of *flux*, or *flow*, of light energy, and it expresses the ability of electromagnetic radiations to produce the sensation of brightness as interpreted by the human eye and the brain.

Luminous Intensity of a Source   The luminous *intensity* of a source *(I)* is defined as the luminous *flux (F)* emitted per unit solid angle. Thus the *unit of source intensity is one lumen per steradian*. This unit is given the name *candela*. Formerly the unit of source intensity was the *candle*. The terms *candle* and *candlepower* (cp) are still encountered, but their use is diminishing rather rapidly.

Since a uniform point source is assumed to radiate equally in all directions throughout a solid angle of 4π steradians, the *total* (visible) *light energy output* F(lumens) from such a source is given by

$$F = 4\pi I \qquad (32.2)$$

Source luminous intensity I is expressed in lumens per steradian. If the older source intensity unit, the *candlepower*, is being used, lumens = $4\pi \times$ cp.

Illumination on a Surface—Illuminance (E)   When luminous flux from a light *source* shines on a surface, the surface is said to be *illuminated*. Assuming uniform illumination over the surface, *illuminance is the luminous flux that strikes the surface per unit area*, or

$$E = \frac{F}{A} \qquad (32.3)$$

The units of illuminance are *lumens per square meter* (called *lux*) in the metric system and *lumens per square foot* [foot-candle (ft-c)] in the English system. Since $1\ m^2 = 10.76\ ft^2$, 1 ft-c = 10.76 lux.

Intensity of illumination (illuminance) on a surface varies inversely as the square of the distance from the source. (This statement is left for you to prove. *Hint:* Use the steradian-solid angle concept.) For *perpendicular incidence*,

$$E = \frac{I}{d^2} \qquad (32.4)$$

where E is illuminance in lux or ft-c.

If the illuminated surface is *not perpendicular* to the flow of luminous energy, the intensity of illumination (illuminance) is given by

$$E = \frac{I \cos \theta}{d^2} \qquad (32.5)$$

where θ is the angle between the line of luminous flow from the source and the normal to the illuminated surface (see Fig. 32.2).

Measuring Source Intensities   The laboratory method for measuring the luminous intensity of an unknown source consists in comparing it with a standard source whose luminous intensity is known. The device used to accomplish the comparison is known as the *bunsen photometer* (Fig. 32.3). It is mounted on an optical bench (see Fig. 32.4) with the standard lamp at one end of the bench and the lamp to be tested at the other. The position of the photometer is adjusted carefully until both sides of the photometer screen are judged to have the same illumination. This judgment is a subjective one, but with practice, considerable skill can be developed. The inverse square law of Eq. (32.4) is then applied, and in terms of Fig. 32.4, $E_1 = I_1/d_1^2$ and $E_2 = I_2/d_2^2$. But $E_1 = E_2$, and therefore

$$\frac{I_1}{I_2} = \frac{d_1^2}{d_2^2} \qquad (32.6)$$

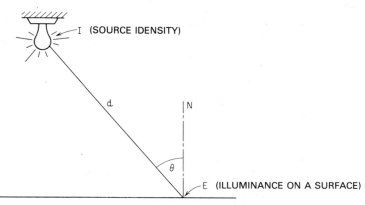

Figure 32.2.   The intensity of illumination (illuminance, E) on a surface that is not perpendicular to the rays of light from the source. $E = I \cos \theta/d^2$.

Figure 32.3. Bunsen-type photometer. The center partition has a star of translucent paper where the brightness (relative illuminances) resulting from the two sources may be compared on a judgmental basis. *(Central Scientific Co.)*

Figure 32.5. Foot-candle meter for measuring the intensity of illumination on a surface. *(Central Scientific Co.)*

The Light-Intensity Meter Direct readings of the intensity of illumination (illuminance) on surfaces will be taken with a light meter (lux meter or foot-candle meter, see Fig. 32.5). A sensitive cell, called a *photronic cell,* in this device converts light energy into electrical energy. These weak electric currents actuate a sensitive electric meter whose dial is calibrated to read directly in lumens/m² (lux) or lumens/ft² (ft-c) or both.

MEASUREMENTS

1. Set the standard lamp, and a 15-W (unknown intensity) lamp, and the bunsen photometer on the optical bench (Fig. 32.6). Cover the apparatus as well as possible to shield out extraneous light. Determine the position of the photometer for equal illumination of both sides of the photometer screen. This will call for great care and deliberation. Approach the equilibrium point from both sides, several times, and use the average of your trial determinations. Record these averages as d_1 and d_2. Test the unknown lamp with the photoelectric foot-candle meter. Choose a distance which will give you a reading of 30 to 40 ft-c (or about 400 lux). Screen out all extraneous light, and be sure to obtain *perpendicular incidence* of the rays on the sensitive cell. Record the exact meter reading and the distance d.

New repeat all these measurements with lamps of 25, 40, 60, 75, 100, and 150 W.

2. Test the formula $E = I \cos \theta / d^2$ by setting up a standard lamp and the light meter as shown in Fig. 32.2. Let the distance and the angle give a reading on the light meter of about 30 ft-c or 300 lux. Measure d and θ, and read E accurately from the light meter.

3. Take the light meter, and check the illuminance (lux or ft-c) on work surfaces in three or four places on the campus, such as

a. Library reading rooms
b. Drafting rooms
c. Medical technology lab
d. Book store
e. Typical classroom
f. Machine shop
g. A faculty office

Complete this tour in about 30 min. Be quiet and unobtrusive. Explain to persons in charge that you are on a class assignment.

Prepare your own Data Sheet for this experiment.

CALCULATIONS

Part 1 Calculate the source intensity of each test lamp from bunsen photometer measurements [Eq. (32.6)] and also from light-meter readings [Eq. (32.4)]. Compare the results in a summary table and compute percent errors, considering the bunsen photometer determinations as the correct values.

Using the photometer values for the source intensity of the lamps you tested, calculate the total light energy (flux) given off by each. Then determine the efficiency of each lamp in *lumens per watt.*

Plot these efficiencies (ordinates) against wattage (ab-

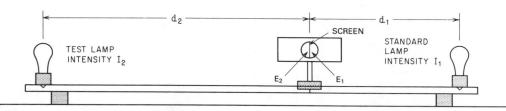

Figure 32.4. Diagram showing conditions for equal illumination on both sides of the photometer screen. The photometer screen is moved along the optical bench until the intensity of illumination on both sides is judged to be equal, that is, $E_1 = E_2$.

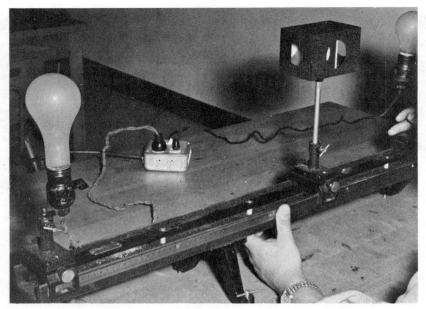

Figure 32.6. Bunsen photometer mounted on an optical bench for a determination of the intensity of an unknown source. The standard lamp is at the right.

scissas), and connect the points with a smooth curve. Label the graph carefully.

Part 2 Verify the formula $E = I \cos \theta / d^2$ from the data recorded in Part 2 of Measurements. Calculate the percent error.

Part 3 There are no calculations for this part, but summarize your findings, and compare them with accepted standards of illumination for similar spaces. (Ask your instructor for a handbook with tables of illumination standards.)

ANALYSIS AND INTERPRETATION

1. Based on collateral reading, write a short history of the development of lighting.

2. Why are ordinary electric lamps called *incandescent* lamps? Why is fluorescent lighting referred to as ''cold cathode'' lighting?

3. Why does fluorescent lighting produce less heat in the room than an equal light flux from incandescent lamps?

4. Write a careful interpretation of the efficiency—wattage curve you prepared from your data.

EXPERIMENT 33

STUDY OF A SPHERICAL MIRROR

PURPOSE

To study image formation by spherical mirrors, and to determine the radius of curvature of a concave mirror and of a convex mirror.

APPARATUS

Optical bench with fittings; concave and convex spherical mirrors; screen, light source with crossed arrows or illuminated wire screen to use as an object; traveling telescope; vernier caliper; spherometer.

INTRODUCTION

In studying image formation with mirrors and lenses it is convenient to think of light *rays* as traveling in straight lines. A light "ray" is considered to be perpendicular to the advancing wavefront of an electromagnetic wave. The wavelengths of electromagnetic waves of visible light are extremely short compared with the dimensions of the lenses and mirrors typically used to form images. It is therefore possible to ignore the *wave properties* of the light and to predict the details of how lenses and mirrors influence light by using a geometric approach involving light *rays* that are assumed to travel in straight lines in any transparent region of constant density. The field of optics in which a geometric, light-ray approach is used to predict how light is influenced by lenses, mirrors, and prisms is called *geometric optics*. In order to study the formation of images by concave and convex spherical mirrors, a geometric statement of the *law of reflection* is needed.

The law of reflection is stated as follows: *The angle of incidence equals the angle of reflection.* The angle of incidence is the angle between the incident ray and an imaginary line *normal* (perpendicular) to the reflecting surface at the point where the reflection occurs. A similar definition applies to the angle of reflection. The incident ray, the reflected ray, and the normal to the reflecting surface all lie in the same plane (Fig. 33.1).

In this experiment we will study image formation by spherical mirrors and will determine the radii of curvature of spherical mirrors by several different methods.

Image formation by spherical mirrors is based on the following assumptions and generalizations (see Fig. 33.2).

1. A light ray which passes through the center of curvature of a spherical mirror on its way to the mirror surface will be *reflected directly back on itself.* (This is a statement of fact.)

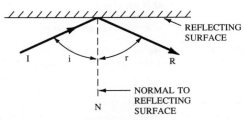

Figure 33.1. Ray diagram showing reflection of a ray of light at a plane surface. The angle of incidence *i* is equal to the angle of reflection *r* for all angles of incidence. The incident ray *I*, the reflected ray *R*, and the normal to the reflecting surface all lie in the same plane.

2. All light rays which approach the mirror in paths parallel to the *optical axis* are reflected through a common point on the optical axis known as the *principal focus*. (An assumption—true only for rays *very close* to the optical axis.)

3. Any light ray which passes through the focal point on its way to the mirror will be reflected parallel to the optical axis. (An assumption—true only for rays very close to the optical axis.)

Since all points on an illuminated object will serve as sources for rays of light, any spherical reflecting surface will reflect the rays which strike it and will bring them together in such a way as to form an *image* of the object, subject to the limitations inherent in the assumptions mentioned above.

For the purposes of the present discussion an image is a *real image* if it can be caught on a screen; and an image is said to be *virtual* if it can only be seen by looking into the mirror. Figure 33.2 shows how rays of light from an object can be reflected by a concave spherical mirror to form a real image of the object.

An equation relating the object distance D_o, the image distance D_i, and the focal length f for spherical mirrors (and also for lenses) is

$$\frac{1}{D_o} + \frac{1}{D_i} = \frac{1}{f} \tag{33.1}$$

or since $R = 2f$,

$$\frac{1}{D_o} + \frac{1}{D_i} = \frac{2}{R} \tag{33.2}$$

Methods of Determining the Radius of Curvature of a Spherical Mirror

1. Concave Mirror If an object is set at a distance from a concave mirror such that a *real image* is formed *at the*

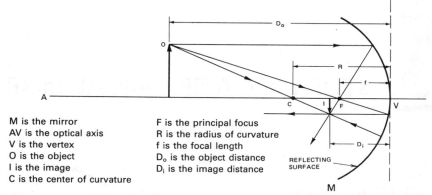

M is the mirror
AV is the optical axis
V is the vertex
O is the object
I is the image
C is the center of curvature

F is the principal focus
R is the radius of curvature
f is the focal length
D_o is the object distance
D_i is the image distance

REFLECTING
SURFACE

Figure 33.2. Ray diagram showing how rays of light from an object are reflected by a concave spherical mirror to form an inverted, reduced, real image of the object. Note how the three representative rays from the arrowhead end of the object intersect at the arrowhead end of the image.

same plane with the object, then $D_o = D_i$, and from Eq. (33.2)

$$\frac{2}{D_o} = \frac{2}{R} \quad \text{and} \quad R = D_o$$

This analysis offers a method of measuring R by a simple determination on the optical bench.

2. Convex Mirror Measurements can be made on an object and on a *virtual image* of the object formed by a convex mirror. The virtual image can be measured by a traveling telescope or comparator.

In Fig. 33.3 let M be the convex mirror surface. AB is a lighted arrow and is the object. $A'B'$ is the virtual image formed behind the mirror. Now, using Eq. (33.2) with proper regard to algebraic signs,

$$\frac{1}{D_o} - \frac{1}{D_i} = -\frac{2}{R} \tag{33.2a}$$

from which

$$D_i = \frac{RD_o}{R + 2D_o} \tag{33.3}$$

Equation (33.2a) can be written as follows:

$$\frac{1}{D_o} + \frac{1}{R} = \frac{1}{D_i} - \frac{1}{R}$$

from which

$$\frac{R + D_o}{D_o} = \frac{R - D_i}{D_i}$$

Transforming,

$$\frac{D_i}{D_o} = \frac{R - D_i}{R + D_o}$$

From the geometry of the diagram,

$$\frac{A'B'}{AB} = \frac{R - D_i}{R + D_o}$$

which, from the previous step, gives

$$\frac{A'B'}{AB} = \frac{D_i}{D_o}$$

Substituting for D_i from Eq. (33.3) above,

$$A'B' = \frac{AB \times R}{R + 2D_o}$$

and

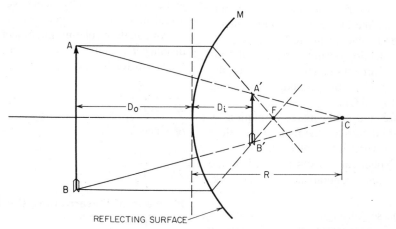

Figure 33.3. Ray diagram showing how rays of light from an object can be reflected by a convex spherical mirror to form an erect, reduced, virtual image of the object. Note how the rays of light from the arrowhead end of the object at A seem, after reflection, to come from the arrowhead end of the virtual image at A'.

Figure 33.4. Laboratory-type spherometer for use in determining the radius of curvature of the surface of a spherical mirror. *(Central Scientific Co.)*

$$R = \frac{2A'B' \times D_o}{AB - A'B'} \tag{33.4}$$

which gives a formula for R in terms of quantities which can be directly measured on the optical bench.

3. Radius of Curvature by Spherometer Measurements The spherometer (Fig. 33.4) consists of a tripod with a micrometer screw at the center. The tips of the three legs determine a plane. The micrometer screw can be lowered below the plane of the leg tips for measurements on concave surfaces, or raised above that plane for measurements on convex surfaces.

Let L be the average distance between the tips of the tripod legs, and d the distance from the plane through the ends of the legs to the point of the screw resting on the mirror surface. (If the spherometer is properly "zeroed," d will be merely the micrometer reading from the zero point.) In these terms, the spherometer formula is

$$R = \frac{d}{2} + \frac{L^2}{6d} \tag{33.5}$$

All measurements for this experiment should be in centimeters.

MEASUREMENTS

Part 1. Image Formation by a Spherical Concave Mirror

Set up the optical bench, object light, concave mirror, and screen as in Fig. 33.5. Determine (and describe completely in your report) the nature of the image formed for six cases, as follows:

Case 1   Object at ∞ (i.e., 5 m or more away)
Case 2   Object outside C
Case 3   Object at C
Case 4   Object between C and F
Case 5   Object at F
Case 6   Object inside F

For Cases 2 and 4 measure accurately D_o and D_i.

Part 2. Radius of Curvature by Forming a Real Image with a Concave Mirror

Keeping the object and the screen in the same plane (screen slightly to one side or slightly above the object),

Figure 33.5. Optical bench arrangement of lamp to illuminate wire-screen object, spherical mirror, and white cardboard screen on which to project the real image formed by the spherical mirror. *(Photo by MSOE Academic Media Services.)*

find an object distance that gives a real image at the same plane as the object. Note that the image is also the same size as the object. When this condition is obtained $R = D_o$.

Part 3. Determination of Radius of Curvature of a Convex Mirror by Measurements on a Virtual Image

Set up the optical bench with the object light at one end and the *convex* mirror surface near the other end. Set up the comparator (traveling telescope) at the object end of the optical bench so that you can sight across the top of the object light at the distant image. Focus the telescope on the distant image, and measure its horizontal length with the comparator. This is $A'B'$ of Eq. (33.4). (See Fig. 33.3.) Measure AB, the object size, with a vernier caliper. Read D_o, the object distance, off the bench. Use two different values of D_o, and collect data for each.

Part 4. Spherometer Measurements

Take measurements on both the concave mirror and the convex mirror with the spherometer, recording data from which R can be calculated, from Eq. (33.5).

CALCULATIONS AND RESULTS

Part 1 Draw neatly and accurately the ray diagrams for the six cases of image formation. Calculate R for the *concave* mirror from the data taken in Cases 2 and 4.

Part 2 No calculations. R was obtained directly. Compare your results with the value given by the instructor and with results from Part 1 above.

Part 3 Calculate R for the *convex* mirror from the measurements taken, using Eq. (33.4). Obtain the mean of your two values.

Part 4 Calculate R for both the concave and the convex mirror from spherometer measurements.

Summarize all results for R in a neat table so that the results from the different methods may be compared.

ANALYSIS AND INTERPRETATION

1. Explain at some length (collateral reading may be required) the errors (aberrations) inherent in image formation by spherical mirrors. How may these errors be minimized? How can some of them be corrected?
2. Discuss at least two applications of concave mirrors in practical or industrial situations. Give a few examples of practical applications of convex mirrors. What is one advantage of a convex mirror over a plane mirror?
3. What can always be said about the image of an object in a convex mirror?
4. In order to obtain a narrow beam of light from a flashlight, where would you place the bulb with respect to the concave reflector surface?
5. What might happen to an object placed at the focal point of a concave mirror "aimed" at the sun? Are there implications here for solar heating?

DATA SHEET

EXPERIMENT 33 ■ STUDY OF A SPHERICAL MIRROR

Part 1. Notes for the Six Cases of Image Formation in a Spherical Mirror (On student-provided sheet.)

Part 2. Notes for the Case $D_o = D_i = R$ (Concave Mirror) (On student-provided sheet.)

Part 3. Determination of R by Measurements on a Virtual Image in a Convex Mirror

Trial	D_o	Length AB (cm)	Length A'B' (cm)
1			
2			

Part 4. Spherometer Measurements

Mirror type	L (cm)	d (cm)
Concave		
Convex		

NOTES, CALCULATIONS, OR SKETCHES

EXPERIMENT 34

INDEX OF REFRACTION

PURPOSE

To study the refraction of light in glass and in water, and to verify Snell's law of refraction.

APPARATUS

Rectangular piece of plate glass; small, rectangular water container with walls of thin, clear plastic; triangular glass prism; pins; ruler and protractor; vernier caliper. *Optional:* Laboratory laser, such as *Metrologic helium–neon laser*.

INTRODUCTION

As light passes from one substance into another, bending of the rays takes place. This bending at the boundary between two transparent substances of different optical properties is called *refraction*. The refracting properties of glass lenses and prisms extend the usefulness and power of the human eye when they are combined in such optical instruments as telescopes, binoculars, microscopes, and cameras.

Refraction (or bending) of light is caused by a change in the speed of light as it leaves one medium and enters another. Light travels fastest in a vacuum (the speed of light $c = 2.998 \times 10^8$ m/s), and only 0.03 percent slower in air. The speed of light in water is about three-fourths that in air. Glass is *optically more dense* than water, the speed in glass being only about two-thirds that in air.

As a ray of light enters an optically dense medium from a less optically dense medium (e.g., entering glass from air), it is bent in a direction which is toward the perpendicular to the surface at the point of entry. In Fig. 34.1, *I* is the *incident* ray (in air) meeting the glass surface at *O*. A portion of the light energy is reflected, of course, but that portion which enters the glass proceeds at a reduced speed and is bent along the path shown by *R* (the *refracted* ray). Let *NN* be the *normal* (or perpendicular) to the boundary surface at *O*. The angle *i* is called the *angle of incidence,* and the angle *r* is called the *angle of refraction.*

The Dutch physicist Snell found that for any two media, no matter what the value of angle *i*, the *ratio of the sine of the angle of incidence to the sine of the angle of refraction is a constant*. By simple geometry (see your physics text) it can be shown that the ratio sin *i*/sin *r* is equal to the ratio of the speeds of light in the two media. *Snell's law* therefore is

$$\frac{\sin i}{\sin r} = \frac{v_1}{v_2} \tag{34.1}$$

where v_1 is the speed of light in the less dense medium (air, for example), and v_2 is the speed of light in the more dense medium (for example, glass or water).

The speed of light v in a transparent medium, the speed of light c in a vacuum, and the *index of refraction* μ of the transparent medium are related by the equation

$$\mu = c/v \tag{34.2}$$

where μ is the *index of refraction* of the transparent medium with respect to a vacuum.

Now, for any two transparent media in which the speeds of light are, respectively v_1 and v_2, $v_1 = c/\mu_1$ and $v_2 = c/\mu_2$. Equation (34.1) can now be rewritten

$$\frac{\sin i}{\sin r} = \frac{\mu_2}{\mu_1} \tag{34.3}$$

or in a more symmetrical form

$$\mu_1 \sin i = \mu_2 \sin r \tag{34.4}$$

where μ_1 is the index of refraction of the transparent medium on the side of the boundary with the incident ray and μ_2 is the index of refraction of the second medium on the side of the boundary with the refracted ray.

MEASUREMENTS

Use the metric system in this experiment.

Prepare your own Data Sheet.

Part 1. Refraction in a Rectangular Glass Plate

Place the rectangular glass plate (Fig. 34.2a) *broad side down* near the center of a sheet of graph paper, and outline

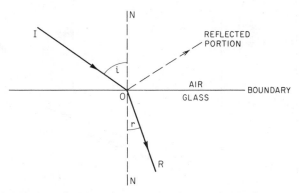

Figure 34.1. Ray diagram showing how an incident ray of light is split into a reflected ray and a refracted ray at an air–glass boundary.

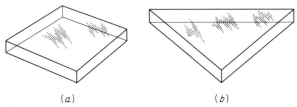

Figure 34.2. (*a*) Rectangular glass plate and (*b*) triangular glass prism for the study of refraction.

the position of the plate with a sharp pencil. The edges of the glass plate should be parallel to the lines of the graph paper.

Insert a pin near the top left corner of the sheet of graph paper. Place a second pin a few centimeters down the page in such a way that a line drawn through the two pins will hit the glass surface at a sharp angle at a point left of the center of the glass surface (see Fig. 34.3)

Now look through the glass horizontally from the other side with your eye down close to the graph paper. (Placing the graph paper on a small box or close to the edge of a table will facilitate this operation.) You should be able to see the shafts of both pins *through the glass*. Move your head from side to side until both pins *appear* to be directly in line. Mark this line of sight by placing a third pin between your eye and the glass plate so that all three pins *seem to be* in line. Finally place a fourth pin near the bottom right edge of the sheet near your eye so that all four pins seem to be in line. Check the alignment *very carefully!*

Remove the glass plate and note that the four pins are *not* actually in a line. With a straight edge draw the incident ray through the locations of pins 1 and 2 to the line which marked the upper edge of the glass plate. Label this intersection as point *A*. Use a fine line. Similarly draw the emergent ray from point *B* through the locations of pins 3 and 4. Now draw the normals N_1 and N_2 and the line *AB*

which represents the *actual* path of the light ray through the glass. With the protractor, measure angles i_1 and r_1 and i_2 and r_2. Measure the horizontal thickness *L* of the glass plate.

Note also the relationship between the incident ray and the emergent ray. Measure carefully the lateral displacement *d* of the ray caused by refraction in the glass plate.

Part 2. Refraction in Water

A small, rectangular water container (about 12 cm long, 6 cm high, and at least 3 cm thick) made with bottom and side walls of thin, clear plastic is to be used (see Fig. 34.4). Fill the container nearly full of water (note temperature) and use it exactly as the glass plate was used in Part 1, employing pins and graph paper to plot the path of a light ray from air on one side, through the water, and out into the air on the other side and on to your eye. Repeat all the measurements of Part 1. *Optional:* Direct a laser beam down on the water surface as indicated in Fig. 34.4. Trace the laser-beam path and water surface on a piece of graph paper held behind the water container, and measure angles as before. *CAUTION! Never look directly down a laser beam!* Severe eye damage could result.

Part 3. Refraction in a Triangular Glass Prism

Place the triangular prism (Fig. 34.2*b*) near the center of a sheet of graph paper with its base parallel to the horizontal lines on the graph paper. Outline its position with a sharp pencil. Place two pins (*Q* and *R* of Fig. 34.5) along the sides which form the prism angle *A*, in such a way that *QR* is parallel to the base of the prism. With the eye at paper level move your head back and forth until the pin *Q* as

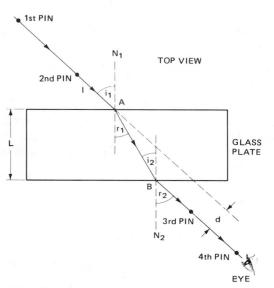

Figure 34.3. Refraction through a rectangular glass plate. This is a vertical view, looking down on the glass plate and the pins.

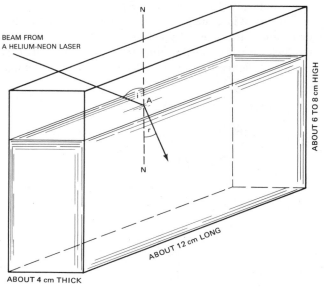

Figure 34.4. Rectangular water container with sides of clear plastic. Available from physics apparatus suppliers, or it can be made up locally.)

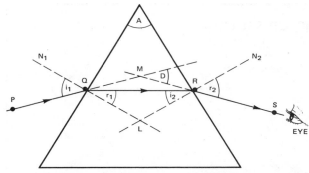

Figure 34.5. Path for a ray of light that is refracted as it enters and as it leaves a triangular glass prism. For the ray path shown, the angle of deviation D will be a minimum.

seen *through the prism* is in line with pin R. With the eye in this position place pin P on the far side of the prism so that all three pins are *apparently* in line. Finally, place pin S on the near side of the prism in line between the eye and point R. Recheck, to be sure that all four pins are (apparently) in the same line of sight.

Remove the prism and draw in lines $PQ, QR,$ and $RS,$ representing the actual path of light from pin P to pin Q and through the prism to pin R and on through pin S to your eye.

Draw in the normals N_1 and N_2 to the prism surfaces, and extend PQ and SR until they intersect at M. The angle D formed by their intersection is the angular amount by which the prism has bent the incident ray PQ. It is called the *angle of deviation*. Measure accurately the prism angle A and the angle of deviation D. Also measure angles $i_1, r_1, i_2,$ and r_2 as accurately as possible.

CALCULATIONS

Part 1. Glass Plate

1. Compute the index of refraction μ of the plate glass from Eq. (34.4). Assume $\mu_{air} = 1.000$.

2. With respect to the emergent ray, what value does the ratio $\sin i_2/\sin r_2$ give? How is this value related to the index of refraction for the glass plate?

3. From the geometry of Fig. 34.3, see if you can derive an equation which will give d in terms of $\mu, L,$ and i_1.

Compare the value of d computed from this equation with the value as actually measured. Use μ as calculated from Step 1 above.

Part 2. Water

From Eq. (34.4) compute the index of refraction of water. Compare your result with the accepted value for water at the same temperature. (*Note:* The plastic sides of the container will introduce some error, but if they are very thin in comparison to the distance through the water, the error will not be large.)

If you used the (optional) laser technique, calculate μ_{water} from these measurements also.

Part 3. Glass Prism

1. Compute the index of refraction μ of the prism glass from Eq. (34.4).

2. Of all possible paths of a ray of light in the prism, the special case of being parallel with the base was chosen for this study. The reason for this choice is that under these conditions the angle of deviation D is a minimum. For this condition, the index of refraction of the prism glass may be computed from this relation:

$$\mu = \frac{\sin \frac{1}{2}(D + A)}{\sin \frac{1}{2}A} \tag{34.5}$$

Compute μ from Eq. (34.5), and compare this value with that obtained from Eq. (34.4). See your physics text for the derivation of Eq. (34.5).

ANALYSIS AND INTERPRETATION

1. Ordinary daylight, "white light," was used in this experiment, and you probably noticed bands of color while looking through the plate glass and the prism. The index of refraction μ (in any medium) *is not a constant* for all colors (wavelengths) of light. Based on collateral reading, explain how it varies with wavelength.

2. By means of a careful diagram and using colored pencils, explain the "dispersion effect" of a prism. Are long wavelengths or short wavelengths bent the most?

3. From the geometry of Fig. 34.5, and assuming that angle D is minimum when QR is parallel to the base, see if you can derive Eq. (34.5).

4. Does an object under water (such as a fish or submerged rock) appear to be deeper or closer to the surface than it really is? Explain your answer by a diagram and the principles of refraction.

5. Discuss the importance of the index of refraction of particular kinds of glass in the manufacture of lenses and prisms for optical instruments.

NOTES, CALCULATIONS, OR SKETCHES

IMAGE FORMATION BY A THIN LENS

PURPOSE

To study image formation by a thin lens, and to determine the lens constants (radii of curvature, refractive index, and focal length).

APPARATUS

Optical bench with several sliding mounts; double convex or plano-convex lens of about 20-cm focal length; illuminated wire screen to use as an object; hooded screen; image screen with magnifier; strips of red glass and blue glass; metal disks, one of the same size as the lens with a central aperture 1 cm in diameter, and one which will cover the central portion of the lens, leaving an outer ring uncovered; spherometer.

INTRODUCTION

Image formation by lenses is perhaps the most frequently encountered phenomenon in the field of optics. Mirrors and prisms are also important components of optical equipment, but lenses are more common, at least in the optical instruments which most of us use. Telescopes, microscopes, spectacles, cameras, projectors, binoculars—in fact our own eyes (see Experiment 36)—are a few of the common optical instruments which depend on the image-forming abilities of lenses.

Lenses which are made of optically denser material than their surroundings, and which are thicker at the center than they are at the edges will bend light rays toward the optical axis (*convergence*), while lenses which are thicker at the edges than they are at the center, bend rays away from the optical axis (*divergence*). Converging lenses are said to be *positive* lenses, and diverging lenses are *negative* lenses.

Simple lenses, like spherical mirrors, produce images that have certain basic defects. One defect—*chromatic aberration*—is caused by the fact that light of different colors is refracted (bent) in varying amounts by the same lens. Consequently, the image of an object which reflects white light is not absolutely sharp, since the shorter (blue) light waves are refracted more sharply than are the longer (red) waves.

Another defect—*spherical aberration*—is the result of the inability of a spherical surface to refract (or reflect) all parallel rays precisely to the same focus. A third defect—*astigmatism*—is noted if the curvature (or the thickness) of a lens is not the same in all planes which pass through the optical axis.

A "thin" lens is defined as one whose thickness is small in comparison to its focal length. The basic lens formula

$$\frac{1}{D_o} + \frac{1}{D_i} = \frac{1}{f} \tag{35.1}$$

applies only to thin lenses. Figure 35.1 shows how the object distance D_o, the image distance D_i, and the focal length f are related for the case where a thin converging lens is used to form a real image of an object.

In order to manufacture lenses which will have a stipulated focal length, a relationship among the *lens constants*

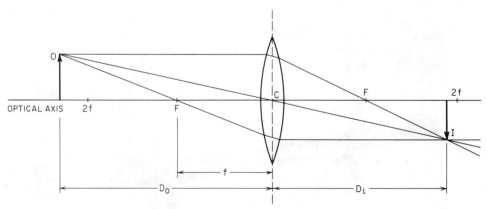

Figure 35.1. Image formation by a thin converging lens. Three representative rays are used—one parallel to the optical axis which is refracted by the lens so that it passes through the principal focus of the lens; one through the center of the lens, which is undeviated; and one through the principal focus on the same side as the object, which travels parallel to the optical axis after refraction by the lens. The intersection of the three rays locates the arrowhead end of the image *I*. (Lens errors neglected.)

must be known. This relationship is expressed in the lens-maker's formula (given here without derivation)

$$\frac{1}{f} = (\mu - 1)\left[\frac{1}{R_1} + \frac{1}{R_2}\right] \qquad \text{(glass in air)} \qquad (35.2)$$

where

f = principal focal length
μ = refractive index of lens glass
R_1 = radius of curvature of first face
R_2 = radius of curvature of second face

(R_1 and R_2 are positive if they represent a curvature which causes convergence and negative if divergence is the result.) Solving Eq. (35.2) for μ for a particular lens gives:

$$\mu = 1 + \frac{R_1 R_2}{f(R_1 + R_2)} \qquad (35.3)$$

This experiment will deal with image formation, lens constants, and certain aberrations of a "thin" converging lens. The mks system of units is used.

MEASUREMENTS

Note: Best results are obtained in a semi-dark room.

Part 1. Determining the Lens Constants

Set up the optical bench (see Fig. 35.2) with four carrier slides, one for the object, one for the lens, one for the hooded image screen, and one for a light source. Fig. 35.3 shows typical equipment for mounting on the carrier slides. An illuminated wire screen may serve as the object. The lens is a double-convex type of ordinary crown glass with a focal length 20 to 30 cm.

With the object screen near one end of the bench and the lens near the center, leave both fixed, and move the image screen until a sharp image is obtained. Measure and record D_o and D_i. Repeat for a different value of D_o that is less than D_i.

Take spherometer measurements on both lens surfaces to determine R_1 and R_2, the radii of curvature of the lens

surfaces. [Review the instructions on the use of the spherometer in Experiment 33. Eq. (33.5) is the spherometer formula.]

Part 2. Focal Length

Point the optical bench toward a distant (bright) window, or toward objects outside the window. The distant bright object is *assumed* to be an "infinite" distance away. Adjust the image screen until the sharpest possible image of the window is obtained. Note that $D_i = f$ for far-away objects (incoming rays essentially parallel).

Part 3. Study of Chromatic Aberration

Since light of different wavelengths is refracted by different amounts, different colors of light are focused at different places along the optical axis by a lens.

Remove the hooded screen, and place an image screen with magnifier (see Fig. 35.3d) in its place. Put a piece of red glass behind the object screen (Fig. 35.3c), and determine a set of values for D_o and D_i. Then replace the red glass with blue glass, leaving D_o unchanged, and redetermine D_i. The optical bench arrangement should be similar to that shown in Fig. 35.4.

Part 4. Study of Spherical Aberration

Use the red glass in order to have monochromatic light and thus not confuse spherical aberration with chromatic aberration.

Cover the lens with a diaphragm which has an aperture of about 1 cm at the center. Determine a set of values for D_o and D_i. Replace this central *aperture* with a central *disk* which allows light to pass through the outer portion of the lens, but *not the central portion*. Leave D_o unchanged, and redetermine D_i.

Part 5. Study of Astigmatism

Leave the red glass and the object screen in place, and use the aperture with central opening. Rotate the lens holder

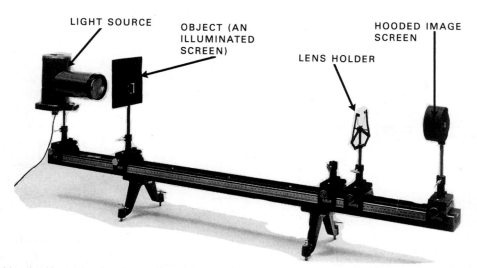

LIGHT SOURCE   OBJECT (AN ILLUMINATED SCREEN)   HOODED IMAGE SCREEN   LENS HOLDER

Figure 35.2. Optical bench with associated equipment. From left: light source, object screen, lens holder, and hooded screen. (*Central Scientific Co.*)

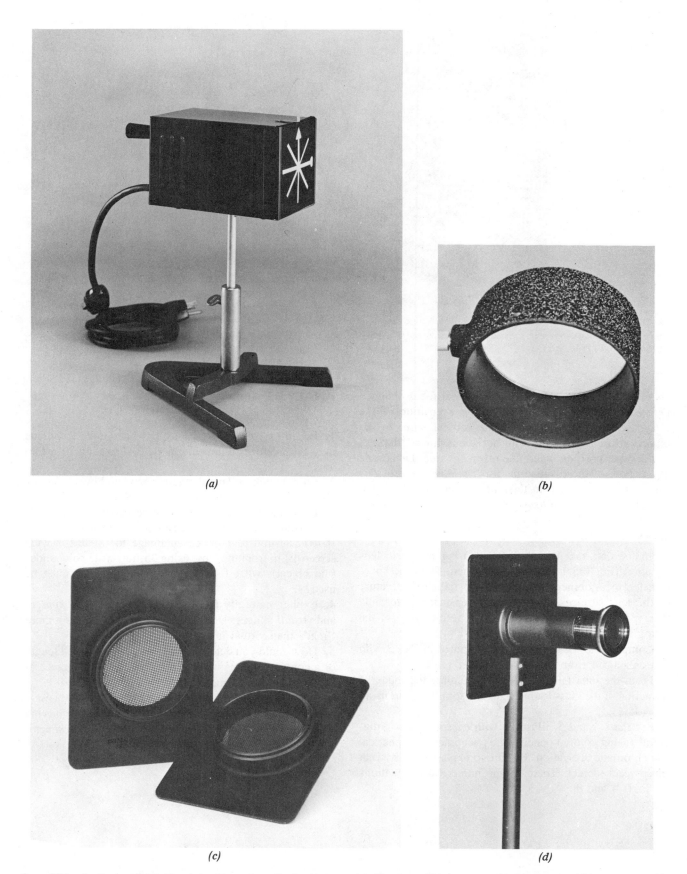

Figure 35.3. Auxiliary equipment for study of image formation by thin lenses: (a) object lamp, (b) image screen, (c) object screens, (d) image screen with magnifier. *(Central Scientific Co.)*

Image Formation by a Thin Lens **181**

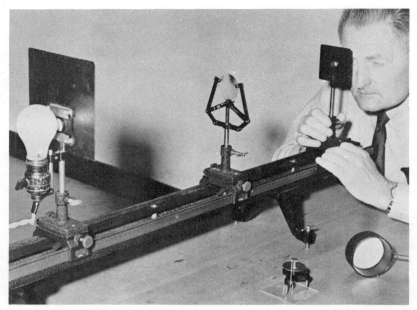

Figure 35.4. Optical bench setup for studies of chromatic and spherical aberration. From left: illuminated object screen, lens holder and lens, image screen with magnifier.

about 45° around a *vertical* axis. The clear focus obtained before will disappear. There will be a position of the image screen magnifier where the vertical wires of the object are in sharp focus with horizontal lines "fuzzy," and another position where the reverse is true. Determine D_i for both of these situations for the *same value* of D_o. This distortion (which in this case is due to a deformed lens) is called *astigmatism*.

CALCULATIONS

From the data taken in Part 1 and using the basic lens equation [Eq. (35.1)], compute f. Then, from the spherometer measurements, calculate the radii of curvature of the lens surfaces. Then use the lensmaker's formula [Eq. (35.2)] to calculate the index of refraction μ of the lens glass for white light.

Compare the focal length as determined in Part 2 with that calculated from data taken in Part 1.

From the data taken in Part 3, calculate the index of refraction of the glass for red light and for blue light, using Eq. (35.3).

For Parts 4 and 5, illustrate with *carefully drawn* diagrams (using colored pencils) the phenomena of spherical and chromatic aberration. Be sure to explain exactly what causes each defect. There are no numerical calculations for Parts 4 and 5.

ANALYSIS AND INTERPRETATION

Answer the following:

1. Based on the results of Part 4, what simple technique may be used to minimize spherical aberration in a thin lens?

2. From text or collateral readings explain how chromatic aberration may be corrected.

3. Draw the ray diagram for the second trial of Part 1. This arrangement is typical of an ordinary slide projector. If it is desired to produce an image to fill a 2-m-wide screen 10 m from the lens, using 35-mm-wide color slides (the object), what focal length projection lens must be used?

4. Explain carefully the difference between *real images* and *virtual images*. Do real images necessarily look more "real" than virtual images?

5. How could you determine the approximate focal length of a converging lens with nothing but the lens, a piece of white paper, and a meter stick?

6. Show by a ray diagram like that of Fig. 35.1 how a converging lens would be used as a simple magnifying glass. Is the magnified image formed by a simple magnifying glass a real or a virtual image?

DATA SHEET

EXPERIMENT 35 ■ IMAGE FORMATION BY A THIN LENS

Part 1. Lens Constants (All measurements in cm)

					Calculate from Eq. (33.5)	
Trial	D_o	D_i	d	L	R_1	R_2
1 ($D_o > D_i$)						
2 ($D_o < D_i$)						

Part 2. Value of Focal Length ($D_o \cong \infty$) $f =$ _____ cm

Part 3. Chromatic Aberration

Trial	Filter	D_o	D_i
1	Red		
2	Blue		

Part 4. Spherical Aberration

Trial	Situation	D_o	D_i
1	Rays through central aperture		
2	Rays through outer edge of lens		

Part 5. Astigmatism

Trial	Situation	D_o	D_i
1	Vertical lines sharp		
2	Horizontal lines sharp		

NOTES, CALCULATIONS, OR SKETCHES

EXPERIMENT 36

THE OPTICS OF THE HUMAN EYE

PURPOSE

To study the optical principles of human eye, and with the aid of an eye model, to identify certain defects in vision and study some of the methods of correcting them.

APPARATUS

Cenco-Ingersoll eye model with associated lenses; object lamp box with radial lines pattern; meter stick; miscellaneous spectacle lenses.

INTRODUCTION

The *normal* eye has a cornea-lens-retina arrangement which brings images of *distant* objects (incoming rays assumed parallel) to a sharp focus on the retina when the eye is fully relaxed. To create sharp retinal images when the object is close to the eye requires a change in the curvature of the crystalline lens. This change is called *accommodation* and is produced by muscles attached to the outer rim of the eye lens (see Fig. 36.1).

Farsightedness (hyperopia) is a defect of vision in which the retinal image is not in sharp focus (it would be formed behind the retina if the retina were not there) because of insufficient convergence of the eye lens. It is corrected by spectacle lenses which supply the needed additional convergence.

Nearsightedness (myopia) is a defect of vision in which

the image is formed in front of the retina as a result of a lens that has too much convergence. The defect is corrected by a lens with just enough divergence to compensate for the excess convergence of the eye lens.

The power of lenses is expressed in *diopters.*

$$\text{Power, in diopters, D} = \frac{1}{\text{focal length, meters}} \qquad (36.1)$$

Converging lenses have positive (+) powers, and diverging lenses negative (−) powers.

Astigmatism is a defect of which either the lens or the retina (or both) are "out of round" or not truly spherical. This defect results in images in certain planes (say the vertical) being sharp and images in other planes being "fuzzy" (See Fig. 36.2). The condition is corrected by the use of auxiliary lenses ground as segments of cylinders. The axis of the cylindrical lens must be so placed that the lens, by its refraction, will correct for the faulty refraction of the eye lens itself.

MEASUREMENTS

Follow through the procedures outlined below, step by step. Record all results and observations carefully. *Make preliminary sketches for each determination.* The sketches should be redone more carefully to submit with your report. Provide your own Data Sheet.

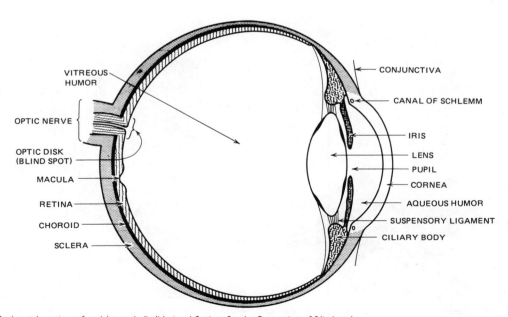

Figure 36.1. Horizontal section of a right eyeball. *(National Society for the Prevention of Blindness)*

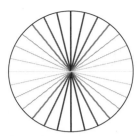

Figure 36.2. A pattern of radial lines all of the same boldness and sharpness constitutes a simple test for astigmatism. Here is a typical view as seen by a person with pronounced astigmatism. The plane in which sharp vision occurs will vary from one astigmatic person to another.

Fill the eye model to within 2 cm of the top with clear water. Keep it level.

1. Accommodation Set up the eye model in a semi-dark room so that it "looks at" the window or some other bright object 6 to 10 m away. With the "retina" in the middle (or normal) position insert the weaker (+7 diopters) of the two crystalline lenses *in the groove just above the lower partition* (in the water). The image of the window (or other distant objects) on which the eye model is pointed should now be in clear focus on the retina. Note the character of the image, whether erect or inverted, its size as compared with the object, etc. Sketch the relationships.

Next use as the object the lamp box with the radially slotted pattern (Fig. 36.3), and place it 35 cm from the "cornea" of the eye model. At this object distance the image will be much blurred until the weaker (+7) crystalline lens has been replaced by the stronger (+20) lens. This process of substituting a thicker lens simulates the process of *accommodation* or focusing on a nearby object, which in the eye, is automatically accomplished by a set of muscles which change the curvature of the eye lens, increasing its convergence.

2. Far- and Nearsightedness With the lamp box placed at 35 cm as before, make the model eye "farsighted" by moving the retina to the forward position in the "eye." Examine the spherical lenses +2.00 and −1.75, and determine whether a converging or diverging lens should be used for correcting the defect and bringing the image to a sharp focus on the retina. Test your conclusion by placing the correcting lens which you select *in front* of the "cornea" of the model eye, and note whether it sharpens the image on the "retina."

Repeat the experiment by moving the retina to the rearmost position (which simulates the nearsighted eye) and then determining what type of auxiliary lens is needed to correct the defect.

3. Effect of Pupil Size—Vision Through a Small Aperture Select any case in which the image is not quite clear, and insert the diaphragm with the 13-mm hole just behind the cornea, and note how the image is improved. This illustrates the fact that a very small aperture in front of the eye lens will sharpen the image, by correcting for spherical aberration. The smaller aperture, of course, reduces the *brightness* of the retinal image, since it lets in less light.

4. Astigmatism Using the lamp box at 35 cm and the retina in the normal eye position, insert *immediately behind* the cornea the cylindrical concave lens marked −5.50, thereby producing astigmatism. (In the human eye, astigmatism is generally caused by abnormalities of the cornea, so in the model a change of cornea would be the logical way of producing astigmatism. This change being impractical, the same purpose is accomplished by the insertion of an additional distorting lens.) By rotating the cylindrical lens slightly, one line alone of the object pattern can be made sharp, the others being blurred. Now place in front of the cornea the correcting convex cylindrical lens marked +1.75, and rotate it until the image is again sharp. Notice the relative directions of the cylindrical axes of the two lenses when the image is sharpest.

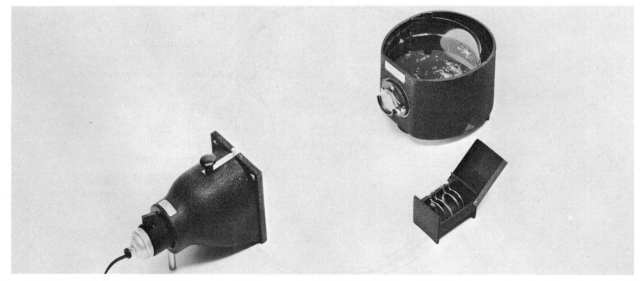

Figure 36.3. Ingersoll eye model (right) with associated test lenses and radially slotted object lamp (left). Note the image showing dimly on the "retina" of the model eye. *(Central Scientific Co.)*

5. Compound Defects Astigmatism is frequently accompanied by farsightedness or nearsightedness. Combine tests 2 and 4, using both a cylindrical and a spherical correcting lens at the same time in front of the cornea. In a spectacle lens the two correcting lenses are combined by the optician in a single compound lens.

6. Vision without Crystalline Lens In some diseases of the eye (cataracts, for example) it is necessary that the crystalline lens be surgically removed. Some vision is still possible, however, by the use of a suitable correcting lens. Remove the simulated eye lens from the model, and place the +7 lens in front of the cornea. *A clear image will be formed if the object is brought very near.*

7. Action of a Simple Magnifier With the eye model retina in the midposition, use the stronger (+20) crystalline lens behind the cornea and set the lamp box at 35 cm. Now introduce the +7 lens as a simple magnifier *in front* of the cornea. For a clear image on the retina the eye model will now have to be moved nearer the object until its distance is only about 1/3 of its former value; the image will accordingly be about three times as large. This shows that a magnifying glass enables the human eye to get close to the object and still see it distinctly. Note that in this connection, for close work, the nearsighted eye has an advantage over the normal eye. With the stronger crystalline lens in place, the lamp at 35 cm, and the retina at normal eye position, note the approximate size of the image. Then make the model eye nearsighted, and move the object until the image is again sharp. It will be considerably larger.

8. Spectacle Lenses With one eye, look in turn through the correcting spherical lenses +2.00 and −1.75 at a printed page. Move the lenses from side to side. Note in which case the print appears to move *with* the lens and in which *against*. How is the effect altered when cylindrical lenses are used and moved parallel to the cylindrical axis? Rotate the cylindrical lenses about an axis perpendicular to the lens. Note the distortion effects.

Now, using your own or a borrowed pair of glasses, test as above, and decide what eye defect(s) they were designed to correct. Note especially if the two lenses are the same, and if not, what the differences are. Look for a correction for astigmatism. Even if it is small, distortion of the image will result as the lens is rotated in its own plane.

A one-diopter lens has a focal length of 1 m; a two-diopter lens has a focal length of 0.5 m; etc. Lens +2 is accordingly of focal length 100/2 or 50 cm. Calculate the focal length of the lens +7, and then measure it approximately, using a distant light as an object. Is the focal length longer in water or in air? In practice, the strength of a lens is most quickly and accurately determined by "neutralizing" it with a lens of the opposite sign. See whether you can, perhaps, neutralize a spectacle lens (yours, or a lab partner's) with one of the lenses in the model set, and thus determine its power.

IN THE LABORATORY REPORT

Your rough sketches should be smoothed up for the report. Draw a neat lettered diagram for *every separate determination*, and show exactly how the optical situation desired is obtained. Indicate, in terms of powers of lenses used, the nature of common eye defects and how they are corrected. Write a separate explanation of each of the eight steps above.

Answer the following:

1. Write a brief history (one-half page) on the early development of spectacles to aid human vision.

2. Does a farsighted person need spectacle lenses which are thicker in the center than at the edges, or vice versa? Why?

3. A farsighted person sees distant objects (i.e., parallel incoming rays) clearly without strain. If he desires to read from a page 50 cm away without the strain of attempted accommodation, what power (diopters) spectacle lenses should be prescribed?

4. Under what conditions are "bifocals" normally specified?

5. Explain, with suitable ray diagrams, exactly how a simple magnifier works.

EXPERIMENT 37
STUDY OF THE REFRACTING TELESCOPE AND PRISM BINOCULARS

PURPOSE

To study refracting-type telescopes, field glasses, and binoculars, and to determine the magnifying power of a simple refracting-type astronomical telescope.

APPARATUS

Optical bench and fittings; a long focal-length converging lens (40 to 60 cm); a short focal-length converging lens (5 to 10 cm); a short focal-length diverging lens (about 15 cm); screen; illuminated object box or an illuminated wire screen to use as an object; a laboratory telescope of the terrestrial type; pair of field glasses which can be disassembled; prism binoculars.

INTRODUCTION

Four types of refracting telescopes will be studied—astronomical telescope, terrestrial telescope, field (or opera) glass, and prism binoculars.

The *astronomical telescope* consists basically of two lenses, a long focal-length converging lens called the *objective,* and a short focal-length converging lens called the *eyepiece.* The eyepiece is used as a simple magnifier to examine the real (inverted) image of the distant object which is formed by the objective lens. The observer's eye thus sees the enlarged virtual image produced by the eyepiece.

The *terrestrial telescope* (Fig. 37.1) uses the same type of objective lens as does the astronomical telescope. An additional positive (converging) lens is used to produce a real inverted image of the original image produced by the objective lens. This second image is therefore *erect,* and it is examined by the eyepiece lens as before, the eye thus seeing an enlarged, virtual erect image.

The *field (opera) glass* (Fig. 37.2) consists of two small Galilean telescopes, one for each eye. Each telescope has the usual long focal-length objective, but the lens used for the eyepiece of each tube is a diverging (negative) lens. This lens is placed at a distance from the objective lens that is equal to the difference between the focal lengths of the two lenses. Rays from the objective lens thus strike the eyepiece lens before they come to a

Figure 37.2. Opera glasses. Each of the two tubes is a simple Galilean telescope. *(Welch Scientific Co.)*

focus, and the real image which would have been formed by the objective lens serves as a *virtual object* for the eyepiece lens. This kind of telescope has a relatively small field of view, but it gives erect images and is convenient to carry because it has the advantage of being shorter than other telescopes.

Prism binoculars (Fig. 37.3) are actually a pair of small telescopes mounted side by side on a frame that is hinged to permit adjustment for the distance between the user's eyes. The objective lenses and the eyepiece lenses, which are converging lenses, would produce an inverted image of the object being viewed were it not for the two totally-reflecting 90° prisms in each barrel. The objective lenses form inverted and horizontally reversed images. One prism reverses the image horizontally and the other inverts it vertically, as shown in Fig. 37.3, so that the net result viewed by the eye through the eyepiece is a magnified, upright, nonreversed image. The double reflections within the prisms effectively extend the length of the path of light, and this allows the use of objective lenses with a focal length considerably longer than the actual body of the instrument. High magnifying power can thus be obtained with a relatively short barrel length.

Binoculars are ordinarily designated as 6 × 30 or 7 × 50, etc. The first number refers to the magnification and the second to the effective diameter (millimeters) of the objective lenses. For example, 6 × 30 binoculars have lenses with an effective diameter of 30 mm and a magnifying power of 6.

The *magnifying power* of a telescope is defined as the ratio of the angle subtended at the eye by the image of an

Figure 37.1. Laboratory telescope, terrestrial type. *(Welch Scientific Co.)*

Study of the Refracting Telescope and Prism Binoculars   **189**

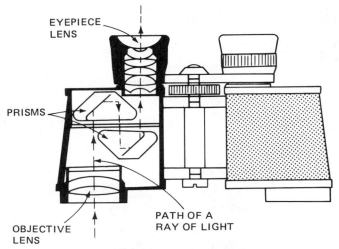

EYEPIECE LENS

PRISMS

OBJECTIVE LENS

PATH OF A RAY OF LIGHT

Figure 37.3. Cutaway view of prism binoculars. Note the positions of the two prisms, their axes at 90° to each other. The prism angles and the index of refraction of the prism glass are chosen to obtain total internal reflection at the 45° faces.

object, to the angle subtended by the object viewed directly. For an astronomical telescope, where the object distance D_o is extremely large, this ratio is essentially equal to the ratio of the focal length of the objective lens (f_o) to the focal length of the eyepiece (f_e). The magnifying power of a telescope is thus given by

$$M = \frac{f_o}{f_e} \qquad (37.1)$$

MEASUREMENTS

1. Determine the focal length of the two converging lenses supplied, using the optical bench and the methods outlined in Experiment 35. Take several trials for each and average the results.

2. The focal length of the negative lens cannot be measured directly. Place it in direct contact with the shorter focal-length positive lens, and determine f for the combination. The focal length of the negative lens can then be calculated from

$$\frac{1}{f} = \frac{1}{f_1} + \frac{1}{f_2} \qquad (37.2)$$

where

f = focal length of the combination
f_1 = focal length of the converging lens
f_2 = focal length of the diverging (*negative*) lens

3. Set up the components of an astronomical telescope on the optical bench. Place the illuminated object box 5 m or more away, exactly on the axis of the optical bench. Adjust the objective (long-focus) lens and the screen until the image of the object box is sharply focused on the screen. Mount the (+) eyepiece (short-focus) lens on the other side of the screen at a distance from the screen equal to its own focal length. Remove the screen, and by readjusting the position of the eyepiece very slightly, obtain a sharp,

virtual, magnified image of the object as you look through the eyepiece.

4. Measure the *magnifying power* of your astronomical telescope while it is on the optical bench by looking through it at a ruler or meter stick some distance away. Look directly at the scale with one eye and through the telescope with the other eye. (This may require a few minutes of practice.) The magnified image from the telescope can thus be superimposed on the unmagnified scale, seen with the unaided eye. By careful counting of the scale divisions on the ruler as seen by the two eyes, you can obtain a ratio which will give you the magnifying power of the telescope.

5. If your instructor approves, partially disassemble the (terrestrial) laboratory telescope (Fig. 37.1), and study its construction. Be careful not to damage the cross-hair arrangement! Diagram the lens arrangement which permits an erect image to be seen. If the eyepiece is fitted with cross hairs, focus the cross hairs sharply. Reassemble the telescope, and determine its magnifying power by the method in Step 4 above.

6. Set up the components of a field (opera) glass on the optical bench, using the same objective lens as in Step 3, but using the divergent (negative) lens for an eyepiece. As before, focus on a screen the image of the illuminated box when the latter is 5 m (or more) away. Now mount the (negative) eyepiece on the bench on the same side of the screen as the objective lens and at a distance from it equal to the difference between the focal lengths of the two lenses. Remove the screen, and make minor adjustments in the position of the eyepiece to enable you to see clearly the *largest possible* erect image through the telescope.

Determine the magnifying power of the field glass, as in Step 4 above.

7. Study an actual pair of field (opera) glasses (Fig. 37.2) if available. Disassemble them if the instructor approves, and study their construction in accordance with the optical-bench study in Step 6. Make a careful diagram of the image-formation process.

8. Study a pair of binoculars (Fig. 37.3), but *do not* disassemble them. Determine their magnifying power and the approximate diameter (mm) of the objective lenses.

DATA SHEET

Prepare your own Data Sheet for this experiment.

CALCULATIONS AND RESULTS

1. Summarize all your results in neat tabular form, one section for the astronomical telescope, one section for the laboratory (terrestrial) telescope, a third for the field (opera) glasses (Galilean telescope), and a fourth for the binoculars.

2. Compare the magnifying power M of the astronomical telescope, obtained from the observations in Step 4, with the value obtained from Eq. (37.1).

3. Draw accurate ray diagrams which will explain the optical principles of the astronomical telescope, the terrestrial telescope, the field glass, and prism binoculars.

ANALYSIS AND INTERPRETATION

1. Based on collateral reading, do the following:

a. Write a paragraph on the early history of the refracting telescope.

b. Explain, using diagrams, how prism binoculars differ from field glasses. What two optical advantages are gained by the introduction of the prisms?

c. What are the essential differences between a reflecting telescope and the refracting type studied in this experiment? What are the advantages of each?

2. Compute the answers to the following:

a. A refracting telescope has an objective lens 15 cm in diameter whose focal length is 90 cm. What must be the focal length of the eyepiece in order to have a magnifying power of 20?

b. The magnifying power of a refracting astronomical telescope is 32. Its overall length from objective to eyepiece when it is focused on a star is 1 m. Find the focal lengths of both the objective lens and the eyepiece.

3. What advantages does a pair of 7 × 50 binoculars have over a pair of 7 × 35 binoculars? Are there any disadvantages?

EXPERIMENT 38

SPECTROSCOPIC ANALYSIS OF LIGHT

PURPOSE

To study the dispersion of white light into a spectrum, and to observe a continuous spectrum and some bright-line spectra.

APPARATUS

Student-type grating spectroscope; spectrum-tube high-voltage power supply; variable rheostat; helium tube; mercury vapor tube; bunsen burner with monochromatic flame attachment; salts of sodium, barium, potassium, and strontium.

INTRODUCTION

White light is made up of many different colors, ranging in a continuous color band from violet through blue, green, yellow, orange, and red. Visible light actually represents a very small part of the total range of electromagnetic radiations. The spread of light waves in the order of their wavelengths, from the deepest red (longest visible waves) to the deepest violet (shortest visible waves) is called the *visible spectrum*. This spectrum includes electromagnetic waves which range in length from about 0.000070 cm down to 0.000040 cm.

In order to avoid the use of long decimal expressions, the wavelength of visible light is usually expressed in units of *nanometers*. One nanometer (nm) is equal to 10^{-9} m (10^{-7} cm). Hence we say that visible light varies in wavelength from about 700 nm (red) to about 400 nm (violet). Occasionally you may encounter the formerly-used unit of light wavelength measurement—the *angstrom* (Å).

One angstrom is equal to 10^{-10} m. Visible light varies in wavelength from about 7000 Å (red) to about 4000 Å (violet). One nm = 10 Å.

When a beam of white light is passed through a prism, it is broken up into the spectrum of colors seen in a rainbow. This process is known as *dispersion,* and it occurs because the velocity with which different colors (wavelengths) of light pass through the prism varies with the wavelength. Since the velocity varies, the index of refraction and the amount of bending also vary. A beam of white light incident on a prism is therefore separated into a spectrum or band of colors as it passes through the prism (see Fig. 38.1 and the "Optical Spectra" color plate found in many college physics textbooks).

A *diffraction grating* also produces spectra by dispersion, and the spectroscope used in this experiment is of the grating type rather than the prism type. (The theory of the diffraction grating is not treated here, as it will have been covered in the lecture classes. The grating spectroscope is specified for the experiment, since this type is of relatively low cost and can therefore be more readily provided for student use.)

Spectrum analysis consists in decomposing a beam of light into its constituent wavelengths and examining the spectrum thus formed. Spectra are of three main types—continuous, bright-line, and absorption. A *continuous spectrum* is produced by light from an incandescent solid or liquid, such as the sun or the tungsten filament of an ordinary electric lamp. A *bright-line spectrum* is produced by the excitation of a gaseous element either by a high-voltage electric discharge or by heating. *Absorption spec-*

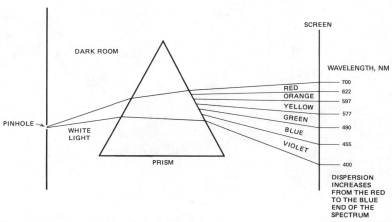

Figure 38.1. Diagram illustrating the dispersion of white light by a prism. Dispersion effect is exaggerated for clarity of diagram.

tra are produced when light from an object which is actually providing a continuous spectrum is seen after passing *through* some cooler absorbing medium.

In this exercise we shall study some examples of *continuous spectra* and *bright-line spectra*.

The principal bright lines in the helium and mercury spectra are given in Tables 38.1 and 38.2.

Table 38.1. Helium Bright Lines

| Wavelength | | |
nm	Å	Color
706.5	7065	red
667.8	6678	red
587.6	5876	yellow
504.8	5048	green
492.2	4922	blue-green
471.3	4713	blue
447.1	4471	deep blue
438.8	4388	blue-violet
402.6	4026	violet

Table 38.2. Mercury Bright Lines

| Wavelength | | |
nm	Å	Color
623.4	6234	red
579.1	5791	yellow
577.0	5770	yellow
546.1	5461	green
491.6	4916	blue-green (aqua)
435.8	4358	blue
407.8	4078	violet
404.7	4047	violet

MEASUREMENTS

1. The Spectrum of the Sun By means of mirrors direct a beam of sunlight into the laboratory and directly onto the slit of the grating spectroscope. (Do not look into a direct beam of light from the sun with the naked eye!) Study the continuous spectrum formed, and record the colors and corresponding wavelengths in sufficient detail so that you will be able to sketch the spectrum with colored pencils in your report. Look carefully for any dark vertical lines that may be visible.

2. The Spectrum of an Incandescent Solid (Tungsten) Place a clear glass (not frosted) electric lamp close to the spectroscope slit and repeat the study of Step 1. Block off all daylight from this test. Record your observations.

3. Bright-Line Spectra from Electric Discharge through Gases Set up the helium gas tube and the spectrum-tube power supply, and connect the apparatus through the variable rheostat to the line voltage. [A high-voltage (30,000 volt) induction coil may be used with a dc source.] Reduce the resistance until the helium tube operates with full glow. (*Careful!* Do not touch the high-voltage contacts of the power supply.) Bring the helium tube to within 2 cm of the slit (see Fig. 38.2), and study the bright-line spectra produced. Record the wavelengths of the light forming the principal lines. (Allow just enough daylight to enter to illuminate the scale.)

Repeat with the mercury vapor tube.

4. Bright-Line Spectra from Particles of Incandescent Salts Set up the bunsen burner with monochromatic flame attachment (Fig. 38.3) a few cm away from the slit of the spectroscope. Heat the barium salt to incandescence and study its spectrum. Record the exact wavelengths of all the bright lines you can see in the spectrum.

Repeat with the other salts.

Supply your own Data Sheet.

RESULTS

There are no numerical calculations for this experiment. For each observation, prepare a neat sketch, using colored pencils, of the spectrum (continuous or bright-line) observed. Label each carefully.

For the gas-discharge spectra, compare the results you

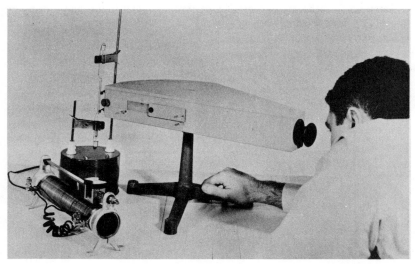

Figure 38.2. Observing bright-line spectra with a grating spectroscope. The helium tube (clamped vertically at left) is shown being operated by a high voltage source and controlling rheostat (at left on table). *(Central Scientific Co.)*

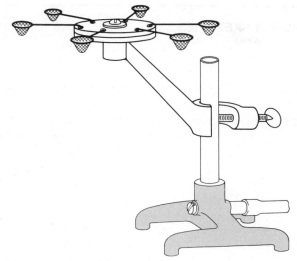

Figure 38.3. Monochromatic flame attachment for bunsen burner. *(Central Scientific Co.)*

observed with the information given in Tables 38.1 and 38.2.

ANALYSIS AND INTERPRETATION

Explain what you have learned about light and color from this experiment. In addition give thoughtful answers to the following questions:

1. In what way is spectrum analysis useful to the astronomer?

2. How would you explain the statement: ''In spectroscopy, each element signs its own name''? In this connection how is spectroscopy used in industry?

3. When you look at a piece of cloth under white light and say it is ''blue,'' exactly what does this mean?

4. If you looked at a red light through a piece of blue glass, what would the effect be?

5. What can you say about the nature of the light energy from a source which gives a bright-line spectrum?

6. If the spectrum of the sun is observed with a precision spectrometer, a series of dark lines (called *Fraunhofer lines*) is observed distributed throughout the otherwise continuous spectrum. You may have observed some of these lines with the instrument used in this experiment. What causes Fraunhofer lines? In what ways are they of importance to astronomers and physicists?

NOTES, CALCULATIONS, OR SKETCHES

39

A STUDY OF POLARIZED LIGHT

PURPOSE

To study some transverse-wave characteristics of light by means of the phenomenon of polarization, and to investigate some practical applications of polarized light.

APPARATUS

Optical bench; screen with adjustable aperture; light source with lens to produce a parallel beam; Polaroid experimental kit (Fig. 39.1); optical bench mount; convex lens of about 30-cm focal length; pieces of cellophane and mica; a simple machine component such as a gear train or a cam, made from clear plastic; small clear-glass medicine bottle; cane and grape sugars; turpentine; liquid crystal display.

INTRODUCTION

According to the wave theory of light, light waves are a form of electromagnetic wave energy which travels through space at a finite velocity, $c = 2.998 \times 10^8$ m/s (186,000 mi/s). Light waves are *transverse* waves, not longitudinal waves. The electric and magnetic vibrations of the wave are perpendicular to each other and also perpendicular to the direction of propagation of the wave. It is assumed that a beam of ordinary light (such as light from the sun or from an incandescent lamp) is made up of millions of such waves, each with its own plane of vibration,

and that there are waves vibrating in all possible planes which intersect in an axis which is the direction of propagation. If such a beam could be observed end-on as in Fig. 39.2, just as many waves should be vibrating in one plane as in any other. This kind of light is said to be *unpolarized light*.

It is possible, by several different means, to produce light in which the vibrations occur in just one of the many possible planes through the axis of propagation. Such light is called *plane-polarized light*.

There are, in general, four methods by which light can be rendered plane-polarized: (1) by reflection from a nonconducting reflector; (2) by double refraction through certain kinds of natural crystals like calcite; (3) by selective absorption in crystals like *tourmaline* and in especially prepared polarizing films; and (4) by the scattering effect of air molecules, water molecules, and dust particles, for example, in the earth's atmosphere. In this experiment we shall be concerned only with polarizing light by selective absorption in special films known as Polaroid.

Polaroid (a trade name for a specified product of the Polaroid Corporation) is made up of a parallel array of long-chain molecules confined in a plastic film. Light that is first plane-polarized in a direction parallel to the long-chain molecules will be absorbed by the Polaroid film. Light that is plane-polarized in a direction perpendicular to the long-chain molecules easily passes through the Polar-

Figure 39.1.   Kit of materials for experimentation with polarized light. *(Bausch and Lomb, Inc.)*

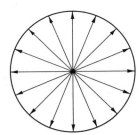

Figure 39.2. End-on view (schematic) of a beam of unpolarized light, indicating that vibration occurs equally in all planes.

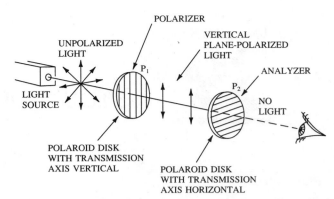

Figure 39.3. A Polaroid film transmits only those components of light waves that are parallel to the transmission axis for the film. No light is transmitted through a pair of (ideal) Polaroid films if their transmission axes are perpendicular to one another.

oid film. The direction perpendicular to the parallel array of long-chain molecules is therefore called the *transmission axis* for the Polaroid film.

The polarizing action of Polaroid films is illustrated in the schematic drawing of Fig. 39.3. A beam of unpolarized light from the source strikes the first Polaroid disk P_1 (called the *polarizer*), oriented so that its transmission axis is vertical. Only those waves with components of vibrations parallel to the transmission axis will pass through; the other vibrations are absorbed. Vertical plane-polarized light is therefore transmitted by the polarizer P_1. If a second Polaroid film P_2 (called the *analyzer*) is now inserted in the path of the light transmitted by P_1 with its transmission axis *horizontal,* the vertical plane-polarized light will be absorbed, and no light will be seen to the right of the second film. If the analyzer is rotated about the optical axis, however, some light will begin to pass, and if it is turned 90° so that its transmission axis is parallel to that of the polarizer, the plane-polarized beam will be transmitted with (nearly) undiminished intensity.

In addition to supporting the theory that light is a transverse wave motion, the phenomenon of polarization has a number of practical applications, which will be studied in this experiment.

MEASUREMENTS

Part 1. Observations of Crossed Polarizers

Set up the optical bench with the universal light source at one end. Adjust its lens system so that it sends out a parallel beam of light (see Fig. 39.4*a*). Place the screen with adjustable aperture close to the light source, and stop down the aperture A to about 2 cm in diameter. Place a screen S at the far end of the bench (Fig. 39.4*b*) and posi-

tion lens L so that a clear image of the illuminated aperture is formed on the screen. The room need not be completely darkened, but the illumination level should be low for satisfactory observations.

Place polarizer P_1 just in front of the aperture with its transmission axis vertical. Rotate it, and note the effect at S. Then place analyzer P_2, as shown, with its axis also vertical. Note the effect at S. Now slowly rotate P_2 through 90°, and note the effect on the screen.

Part 2. Effects of Cellophane and Mica on Plane-Polarized Light

Leave the Polaroids set for extinction. Insert a piece of mica between the Polaroids, and observe the effects of

(*a*)

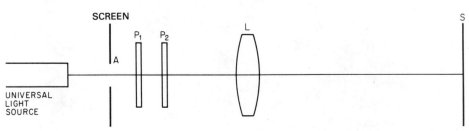

(*b*)

Figure 39.4. (*a*) Universal light source. (*Central Scientific Co.*) (*b*) Arrangement of equipment on an optical bench for polarized light observations.

twisting, bending, and rotating it on the beam of polarized light. If effects are not easily visible on the screen, remove it and look through the analyzer.

Make the same observations with a piece of cellophane, folding it to several thicknesses and bending and twisting it.

Part 3. Strains in Transparent Materials

Homogeneous transparent materials like glass and clear plastic do not ordinarily affect light which passes through them. When they are subjected to unusual external or internal stresses, however, the strained regions undergo certain changes in their optical properties. The changes result in optical behavior which is characteristic of some crystals rather than of amorphous materials like glass. The visible effect, as seen by polarized light, is that of varicolored striations at the points of strain.

Place one of the strain samples (from the Polaroid kit, or the gear or cam models made up locally) in the space between the two Polaroids. Move the lens *L* as necessary to focus the image of the strain sample on the screen. Set the Polaroids for extinction. Now put the sample under *stress* with your fingers, clamps, pliers, or by any suitable method. This stress causes *strain* in the material of the sample. Note carefully the image on the screen. Sketch the strain sample and the strain points, as shown up by the test.

Repeat with the other samples, making careful sketches of the strain effects.

This application of polarized light is used in industry to detect strains in optical glass, in mounted lenses, in articles made of clear plastic, and in expensive laboratory glassware. By using transparent scale models of machine and structural components, strain points under operating conditions can be predicted in advance, and any indicated design changes can be made prior to quantity production.

Part 4. Rotation of the Plane of Polarization by Optically Active Substances

a. Sugar Solutions and Turpentine Certain sugar solutions and turpentine are common substances which possess the ability to rotate the plane of polarization of polarized light which passes through them. For a standard length of optical path through the solution, the degree of rotation caused by the passage of plane-polarized light through a sugar solution is a measure of the concentration of the solution.

Place the empty medicine bottle between the Polaroids, and set them for extinction. Then fill the bottle with tap water, and see if any change occurs in the transmitted light pattern. Next fill the bottle with a cane sugar solution of a concentration equivalent to about 3 tsp of sugar in a cup of water. Some light should now be observed on the screen. Determine the angle through which the plane-polarized beam has been rotated by noting how far (in degrees) the analyzer must be turned to extinguish the light again. In which direction does cane sugar rotate the beam?

Repeat with a cane-sugar solution whose strength is that of near saturation.

Repeat with similar strength solutions of grape sugar. Note the direction of rotation of the polarized beam.

Repeat with turpentine.

b. Liquid Crystal Displays The liquid crystal displays (LCDs) used in digital watches, electronic calculators, digital multimeters, and many other applications depend on the ability of molecular arrangements in a liquid crystal to rotate the plane of polarization of plane-polarized light through an angle of 90°. Figure 39.5 illustrates the six-layer sandwich of materials that make up a typical liquid crystal display.

Note that light striking an LCD first passes through a polarizing sheet. Therefore, if a sheet of Polaroid film is held in front of an LCD and oriented so that its transmission axis is perpendicular to the transmission axis of the first polarizing sheet in the LCD, essentially no light will be transmitted through the liquid crystal to the reflecting surface at the back of the LCD.

Hold a Polaroid film in front of the LCD on a digital watch or on a laboratory instrument and rotate it to find the orientation that makes the entire display of the LCD appear quite dark.

DATA SHEET

Provide your own Data Sheet for this experiment. Take down all notes and measurements necessary to enable you to write a good report. The Data Sheet must be neat and well organized.

RESULTS

There are no numerical calculations. Present your results under the four headings listed under Measurements. Describe exactly what you learned, and use sketches freely to clarify the report.

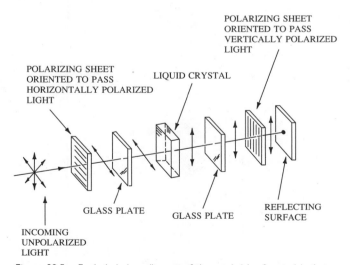

Figure 39.5. Exploded view diagram of the sandwich of materials that make up a liquid crystal display (LCD).

ANALYSIS AND INTERPRETATIONS

Answer the following:

1. Explain how the phenomenon of polarization supports the theory that light is a transverse-wave motion.

2. From collateral reading give some *actual examples* of the industrial uses of polarized light under the following headings:

a. Strains in transparent materials

b. Rotation of plane of polarized light by optically active substances. With respect to sugars, what is the origin of the terms *dextrose* and *levulose?*

3. Explain (collateral reading will be necessary) how "Polaroid" sunglasses reduce glare from sunlight and water surfaces.

4. Explain why the noonday sky and deep bodies of clean water are blue.

5. Why do "Polaroid" sunglasses ease the glare from highway surfaces for the auto driver and assist the trout fisherman in seeing fish below the water surface?

EXPERIMENT 40

INTERFERENCE OF LIGHT, USING A LASER BEAM

PURPOSE

To study the interference of light by a modified double-slit method, and to calculate the wavelength of laser light.

APPARATUS

Helium-neon continuous-wave gas laser (*Metrologic* or equivalent); interference slit-film demonstrator kit (*Cornell* or equivalent); viewing screen about 1 m wide; metric ruler, meter stick; magnifier; stands and clamps.

INTRODUCTION

When the paths of two waves cross, interference between the waves results. If the waves are of the same frequency and *in phase,* the resulting wave will have an amplitude greater than that of either contributing wave (constructive interference). (See Fig. 40.1.) If the waves are of the same frequency but 180° *out of phase,* the amplitude of the resulting wave is diminished (destructive interference) (see Fig. 40.2). If both waves start with equal amplitudes, the in-phase waves yield a doubling of amplitude, while the 180° out-of-phase waves nullify each other at every point and there is complete cancellation of the wave energy. Some examples of destructive interference are:

1. Water waves Two vibrating floats near each other, being forcibly oscillated at the same frequency and amplitude, will produce lines of still water (i.e. no waves).
2. Sound waves Slowly rotating a tuning fork near your ear while it is vibrating will produce a diminished sound at certain orientations, and a louder sound at other orientations.
3. Light waves A light source passing through two closely spaced slits will produce alternate bright and dark lines on a nearby screen.

It is this third phenomenon that this experiment investigates, as a beam of light from a laser passes through closely spaced transparent slits in an otherwise opaque screen.

In 1800, Thomas Young first demonstrated the interference of light waves by passing filtered sunlight through two narrow slits and observing the resulting pattern on a screen. A pattern such as he observed is shown in Fig. 40.3. Although a single slit produced a continuous spread of light (as observed by the unaided eye), two slits produced "bands" on the screen where no light was found. This *interference pattern,* and Young's other findings, were instrumental in establishing the wave theory of light.

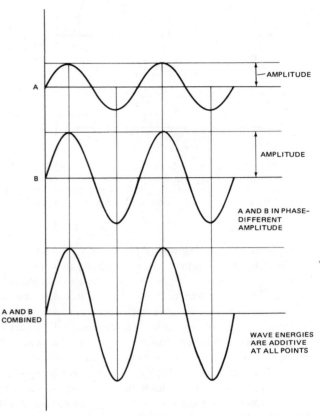

Figure 40.1. Constructive interference. Waves A and B are in phase but of different amplitudes. Their cumulative result is a wave of the same frequency and of larger amplitude than either A or B. For light waves this means greater light intensity, or increased brightness.

The spacing between the bright and dark lines can be found from considerations of geometry. The geometrical derivation is not given here, but it can be found in many standard college physics texts. The relationship is given by

$$\lambda = \frac{xd}{D} \tag{40.1}$$

where

λ = wavelength of light used
x = distance between adjacent bright lines
d = slit separation distance, center-to-center
D = slit-to-screen distance

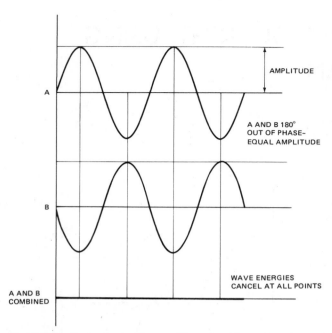

Figure 40.2. Destructive interference. Waves A and B are of equal amplitude and are 180° out of phase. Their cumulative result is a wave of zero amplitude at all points. For light waves this means complete darkness. The wave energies cancel each other out at all points along the waves.

By measuring the geometrical factors x, d, and D, the wavelength of the light λ can be readily calculated.

In this experiment, a gas laser will be used to provide monochromatic, coherent light. *Monochromatic* means light of a single frequency (color). *Coherent waves* travel in phase and spread very little. A laser beam will travel for long distances with very little spreading of the beam. A number of slit patterns in the slit-film sheet will be used. Each one differs either in slit spacing or in number of slits. Since the diameter of the laser beam is quite small, you will see points of light on the screen rather than bright lines of light.

Safety Note: Never look directly into the laser beam! Although the laboratory laser is of low power and will not burn your skin, the beam can do serious damage to the

retina of the eye. When measuring your x-dot spacings, *do not look back at the beam!*

MEASUREMENTS

Provide your own Data Sheet for this experiment.)

Mount the slit film with clamp and stand and place it about 5 m from the viewing screen. Carefully measure the distance D from slits to screen and record. Adjust the laser until it is in front of the one-slit pattern (see Fig. 40.4).

Record your observations of the pattern on the screen obtained from just one slit. Then change the viewing pattern to, in turn, 2, 4, 10, and 80 slits, examining the patterns of the viewing screen and recording your observations each time.

What happens to the resolution (i.e., clarity and sharpness of the image) as a greater number of slits are used? Select the pattern that gives the best resolution on the viewing screen and measure (with the metric ruler, and using the magnifying glass as needed) the distance between adjacent dots on the screen. Are they of equal spacing? If so, divide the total distance between the end dots by the number of intervals to find the average distance x of Eq. (40.1). If they *are not equally spaced*, use the distance between the brightest dot and the one next to it as the value for x.

Observe the brightness of each dot and record your observations. Are the dots of equal brightness? Is the brightness constant *within each dot?*

Repeat the measurement of the distance x for the following additional cases: (1) Any slit pattern of different slit spacing, d, with D the same as before; (2) the same slit pattern as (1), but using a different slit-to-screen distance D.

CALCULATIONS AND RESULTS

In the slit-film kit instruction book you will find slit spacing data and reduction-factor information. Use this information to calculate the actual spacing between slits. Remember that d represents the center-to-center slit separation.

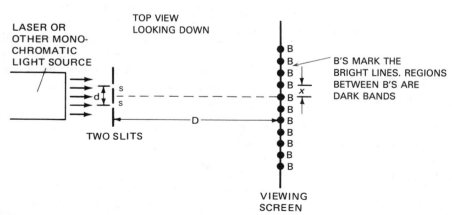

Figure 40.3. Sketch illustrating Young's interference experiment. Constructive interference of light from the two slits (S··S) results in equally spaced bright spots B, a distance x apart, on a viewing screen located a distance D away from the two illuminated slits of slit-spacing d.

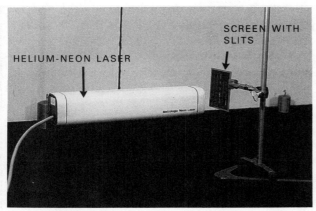

Figure 40.4. Helium-neon (He-Ne) laser shown mounted immediately behind the opaque film with the transparent slits. Viewing screen is off to the right.

Use Eq. (40.1) to calculate the wavelength of the laser light for each set of data. Find out from the instructor the actual wavelength of the laser light used. Calculate the percent error from this value, of the results you obtained from each set of data.

ANALYSIS AND INTERPRETATION

1. What effect does the number of slits used have on the sharpness of the interference pattern?

2. Are your calculated results for the wavelength of the laser light more accurate for larger or for smaller slit spacings? For larger or smaller slit-to-screen distances? Explain.

3. Why does the dot brightness decrease on both sides of the central dot?

4. You often see the terms "interference" (of light) and "diffraction" mentioned together. What is *diffraction?* Has diffraction entered into this experiment in any way?

5. Based on some reference reading, and using diagrams, explain (in an elementary way) the theory and operation of a laser. What are some of the uses of laser beams in industry, medicine, communications, and the military?

6. A pair of parallel transparent slits in an opaque screen are separated by a distance of 0.20 mm and are illuminated by laser light of wavelength 600 nm. For what slit-to-screen distance D will the distance x between adjacent bright spots in the interference pattern be equal to 50 times the slit-separation distance?

NOTES, CALCULATIONS, OR SKETCHES

PART SIX ■ MAGNETISM AND ELECTRICITY

EXPERIMENT 41

MAGNETIC AND ELECTRIC FIELDS

PURPOSE

To study the nature of magnetic and electric fields.

APPARATUS

Bar magnets, horseshoe magnet, slotted magnet board, iron filings; several small magnetic compasses (1-cm diameter); one larger compass (3–5-cm diameter); Robison spherical-ended magnet (or other apparatus providing isolated N and S poles about 20 cm apart); electric field-plotting apparatus; sensitive galvanometer; source of 12-V dc; plotting paper (large sheets).

INTRODUCTION

The phenomenon of magnetism has been observed and studied for centuries. For about 2000 years the only practical application of magnetism was in the development of navigation. Then, beginning with the studies of Coulomb and of Oersted (1820), knowledge of magnetism and magnetic fields has developed, until today, applications of magnetism are basic to such processes and equipment as electric motors and generators, the operation of communications and sound systems, the operation of cyclotrons and other "atom smashers," experiments in controlled thermonuclear fusion, research in low-temperature physics *(cryogenics),* and magnetic resonance imaging in medicine, to name but a few.

The most easily observed attribute of a magnet is that it exerts forces on other magnets or on pieces of iron. These forces occur not only on contact, but within a region or "field" surrounding the magnetic pole. They may be forces of attraction or forces of repulsion, depending on the particular circumstances.

Man's earliest experiences and observations with electricity dealt with electricity at rest, i.e., with *electric charges.* It was Thales (about 650 B.C.) who first observed that a piece of amber rubbed with wool would exert forces of attraction on particles of chaff and other nonconducting materials. *Electrostatics,* too, was nothing more than a curiosity for centuries. Beginning, however, with the investigations of Franklin and Gilbert, an increasingly rapid development of knowledge about electricity occurred. One of the important phenomena associated with electric charges is that there is surrounding the charge a "field" of force which is similar in some ways to the field of force surrounding a magnet but in other ways is different in its properties.

The concept or idea of a *field* is basic to the development of our knowledge of magnetic and electric phenomena. Magnetic forces have their origin at the magnet, and electric forces have their origin at the point of electric charge. Surrounding a magnet or an electric charge there are, however, areas (or spaces) in which "action-at-a-distance" forces can act. These regions or spaces in which magnetic and electric forces are present and detectable are called *fields.* The field concept is of extreme importance in understanding many of the phenomena of *modern physics.* From a study of the forces of interaction between planets and stars (the *macrocosm*) to the investigation of forces between atomic particles (the *microcosm*), the idea of a *field of force* is a part of the central core of scientific knowledge.

This experiment will help you to understand some of the properties and effects of magnetic and electric fields.

MEASUREMENTS

Plot all observations carefully. Give each plot a title.

Part 1. Magnetic Fields

1. Study of Magnetic Fields as Indicated by the Action of Iron Filings

a. Set up the bar magnets in the slotted board (Fig. 41.1) so that two N poles are about 2 in. apart. Cover with a sheet of paper and sprinkle on iron filings. Sketch the "lines of force" which make up the magnetic field as indicated by the configuration of the iron filings.

b. Repeat with an N pole—S pole arrangement of the bar magnets.

c. Repeat with a horseshoe magnet.

2. Plotting a Magnetic Field Produced by the Interaction of the Field of a Magnet with the Earth's Magnetic Field

Fasten a Robison spherical-ended magnet to the underside of a wood table, well away from any pipes or other iron.

Figure 41.1. Slotted board with bar magnet (left). U-shaped (horseshoe) magnet (right).

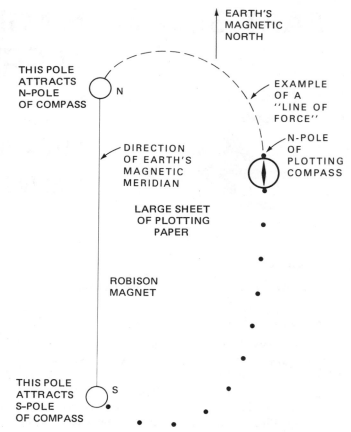

Figure 41.2. Method of tracing a line of force between magnetic poles of a Robison (spherical-ended) magnet with a small compass needle. The field of the Robison magnet interacts with the earth's magnetic field.

Its axis should be parallel to the magnetic meridian of the earth as determined with the larger compass. (In other words, the pole of the Robison magnet which attracts the N pole of a compass needle should be pointed to the earth's *magnetic north pole*.) Always hold compasses well away from any iron or steel body.

Over the magnet, and on top of the table, fasten a large sheet (18 by 24 in.) of plotting paper. Mark the direction of the earth's magnetic meridian on the paper, and locate points N and S on the paper directly above the poles of the Robison magnet. Draw circles around these points with diameters equal to those of the magnet spheres.

Trace three or four force lines from circle N to circle S *on each side* of the magnet, using a small plotting compass (see Fig. 41.2). Be sure to locate several *singular points* (points at which the compass needle quivers, assuming no definite direction). At these points the resultant field is zero, that is, the earth's magnetic field just balances that of the Robison magnet. Label carefully all features of your plot.

Part 2. Electric Fields

Set up the electric field-mapping apparatus (Fig. 41.3), and connect it to a 12-V dc source. Connect one wire from a sensitive galvanometer to the U-probe, and the other wire to the farthest left of the seven sockets of the series resistors at the top of the board. Clamp a piece of plotting paper on top of the board.

First, use the field plate which simulates "two points within a field," and with the template which accompanies the apparatus, draw in the location of the two "charged

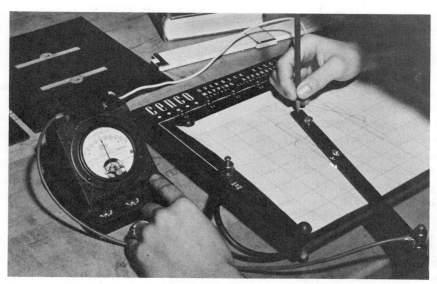

Figure 41.3. Plotting an equipotential line between the plates of a simulated capacitor, using the electric-field mapping apparatus. *(Central Scientific Co.)*

206 Experiment 41

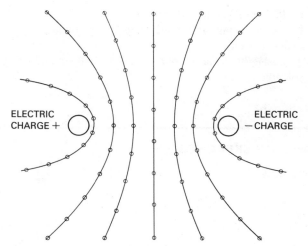

Figure 41.4. Sample plot of equipotential lines when a field plate simulating "two charged points" is used.

ELECTRIC CHARGE +

ELECTRIC − CHARGE

points'' on the plotting paper. Then, with the probe, locate a series of equipotential points (indicated by zero deflection of the galvanometer) around the left-hand charged point. Move the galvanometer lead right to the next socket, and plot another equipotential line. Continue until you have a line for each of the seven possible galvanometer lead positions. Your plot should resemble Figure 41.4.

Repeat the entire process with the field plate which simulates a *parallel plate capacitor*. The plot for this exercise should resemble that in Fig. 41.5. Note that these are lines of equipotential. The electric *force field* will be represented by a set of lines that are everywhere perpendicular to the equipotential lines.

DATA SHEET

None. Observations are to be recorded on the plotting sheets.

PRESENTATION OF RESULTS

There are no numerical calculations for this experiment. The plots and sketches you have made, together with the interpretations which you will make of them, constitute the results of the experiment.

Part 1. Magnetic Fields

1. Make careful sketches of the field (as indicated by the patterns of iron filings) between the bar magnets (N to N and N to S) and between the poles of the horseshoe magnet. Indicate the ''direction'' of the lines of force by arrowheads. Explain how magnetic lines of force seem to act like ''stretched rubber bands.''
2. For the Robison magnet–earth's field plot, connect the plotted points by curved lines. These are *lines of force*. Indicate their direction, N to S.

In red pencil draw in a set of lines (three or four for each spherical magnet pole) which are everywhere perpendicular to the lines of force. These will be curved lines also, and they are called *equipotential lines*.

Be sure to indicate the location of several *singular points*.

Label everything and give the plot a title. It must be neat!

Part 2. Electric Fields

For both plots (the two charged points, and the simulated parallel plate capacitor) connect the plotted points with curved lines. These are *lines of equipotential*. With a red pencil draw in a set of lines everywhere perpendicular to these equipotential lines. This latter set represents *lines of force* in the *electric field*.

Label everything and give the plots a title.

ANALYSIS AND INTERPRETATION

Summarize what you have learned about magnetic and electric fields. Answer the following:

1. How are energy and work related to force fields?
2. On what kinds of substances can magnetic fields exert forces? Electric fields?
3. Explain in some detail one practical application of magnetic fields and one of electric fields.
4. What problems for space exploration are posed by the presence of magnetic and electric fields in space and around celestial bodies?
5. Based on collateral reading, write a half-page report on

Figure 41.5. Sample plot of equipotential lines when the "parallel plate capacitor" field plate is used. *(Central Scientific Co.)*

the use of strong magnetic fields in any one of the following applications:

a. research on thermonuclear fusion
b. cyclotron and other ''high-energy physics'' operations
c. cryogenic research
d. television and radar
e. levitation, as in rail transport
f. medical diagnosis, using magnetic resonance imaging (MRI) systems
g. possibilities for energy storage in ''superconducting'' solenoids

---

NOTES, CALCULATIONS, OR SKETCHES

EXPERIMENT 42

MEASURING ELECTRICAL RESISTANCE—OHM'S LAW

PURPOSE

To study some methods of measuring and calculating electrical resistance, and to determine the resistivity of copper and German silver (or nichrome).

APPARATUS

Voltmeter (0 to 5 V); ammeter (0 to 1 amp); switch; bank of test coils; slide-wire Wheatstone bridge; galvanometer; source of low-voltage (not over 2 V) direct current; resistance decade box; ohmmeter; special test coil (length known only to instructor); micrometer caliper; wire for connections.

INTRODUCTION

Resistance to Electron Flow—Ohm's Law

Ohm's law states that a simple relationship exists between the current in a circuit, the resistance of the circuit, and the electromotive force (or voltage) applied. If current (I) is measured in amperes (A), resistance in ohms (Ω), denoted by R; and the electromotive force in volts, denoted by V, then Ohm's law may be written

$$I = \frac{V}{R} \tag{42.1}$$

In words:

$$\text{Amperes} = \frac{\text{volts}}{\text{ohms}}$$

Resistivity   The resistance of a conductor is directly proportional to its length L, and inversely proportional to its cross section area A. It can be calculated from the relation

$$R = \rho \frac{L}{A} \tag{42.2}$$

where ρ (the Greek letter *rho*) is a constant of the conductor material, called its *resistivity*. In calculating resistances of wires in the *English system* it is a common practice to express length in feet and diameter in *mils,* a mil being defined as 0.001 in. The cross-sectional area of the wire is then expressed in *circular mils* in accordance with the following relation:

Area (in circular mils) = [diameter (in mils)]^2

In the English system, for a round wire, resistivity is equal numerically to the resistance of a wire 1 circular-mil in cross-sectional area and 1 ft long. *Resistivity* is, then, expressed in *ohms-circular mil/ft* ($\Omega \cdot$ cmil/ft). If we let L stand for wire length (ft), and D for diameter (mils), then the total resistance (R) of the wire is

$$R = \frac{\rho L}{D^2} \tag{42.2'}$$

In the SI-metric system, the resistivity of a substance is equal numerically to the resistance of a sample of the material 1 m long and 1 m^2 in cross-sectional area. The unit of resistivity in mks units is therefore the ohm-meter (Ω-m). For the metric system A is the cross-sectional area in *square meters* (m^2).

Both systems of measurement will be used in this experiment.

Resistance Measurement by the Voltmeter-Ammeter Method   A set of test coils consisting of different lengths and sizes of wire of two different materials is suggested (see Fig. 42.1). Electrical connections are made in accordance with the circuit diagram shown in Fig. 42.2. Note that I is the total current in the circuit and V is the voltage drop across the resistance R. The ammeter is *in series* with the unknown resistance and the voltmeter is connected *across (in parallel with)* the unknown resistance. When the switch K is closed, I and V are read from the meters. The resistance can then be calculated from Ohm's law [Eq. (42.1)]. It should be pointed out, however, that the voltmeter, being connected in parallel in the circuit, also carries a small current, and consequently, this method of resistance measurement does not have a high degree of precision.

Take readings of I and V for each test coil.

Resistance Measurement by the Wheatstone Bridge   The Wheatstone bridge measures the resistance of unknown resistors by providing a balance between two parallel circuits. One branch of the circuit is a uniform resistance wire of known length, and the other branch contains the unknown resistance and a known resistance, the latter usually a setting on the dials of a resistance box.

In circuit diagram form (see Fig. 42.3) let AB be the uniform wire, R a known resistance dialed on a resistance box, and R_x the resistance which is to be measured. G is a sensitive galvanometer and C is a sliding contact which may be touched anywhere along AB. Fig. 42.4 shows how the apparatus may be connected.

When switch K (Fig. 42.3) is closed, current flows to A [conventional direction of current ($+$ to $-$) is assumed], then divides and flows to B by the two paths ADB and ACB. If the values of R and R_x and the lengths of L_1 and L_2 are so chosen that the galvanometer shows zero deflection, points D and C will be at the same potential. Under these conditions

$$\frac{R_x}{R} = \frac{L_1}{L_2}$$

and

$$R_x = \frac{RL_1}{L_2} \qquad (42.3)$$

MEASUREMENTS

Part 1. Resistance Measurement by the Voltmeter-Ammeter Method

Set up the apparatus as in Figs. 42.1 and 42.2, and take simultaneous voltmeter and ammeter readings on each of the five test coils on the board. Measure the diameter of each wire with the micrometer. Note the given lengths of the wires. Record all measurements on the Data Sheet, in both mks and English units.

(a)

Part 2. Resistance Measurement by the Wheatstone Bridge Method

Place the special test coil provided by your instructor as the unknown resistance R_x of a Wheatstone bridge circuit as diagrammed in Fig. 42.3 and as illustrated in Fig. 42.4. Adjust the known resistance R (resistance box) and the point C along the wire until the galvanometer G reads zero. Take measurements of L_1 and L_2 when R (the resistance set on the dials of the resistance box) results in zero galvanometer deflection. Take two more trials, using two different values of R. (The value of R should be adjusted so that the point C will fall within the middle third of the wire AB when the galvanometer reads zero.)

Record the American Wire Gage (AWG) (and the diameter) of the wire on the test coil.

CALCULATIONS AND RESULTS

From the measurements taken in Part 1 and using Ohm's law, calculate the resistance of each of the five test coils.

Using these values of the resistance and the basic resistivity equation [Eqs. (42.2) and (42.2′)], calculate the resistivity of copper (four trials, and average the results) and the resistivity of German silver (or nichrome) in both mks and English units. Compare with values from published tables and calculate the percent error of your experimental results.

From the measurements taken in Part 2 with the Wheatstone bridge, calculate [using Eq. (42.3)] the resistance of the special test coil for all three trials. Average the results and compare your average value with the value obtained using a direct-reading ohmmeter. Determine the diameter of the wire from the AWG table (or measure its diameter), and look up the resistivity of copper in the tables. With these data, and using Eq. (42.2), determine the length of wire (in feet and in meters) on the special test coil. Compare your calculated length with the actual length, as given by the instructor.

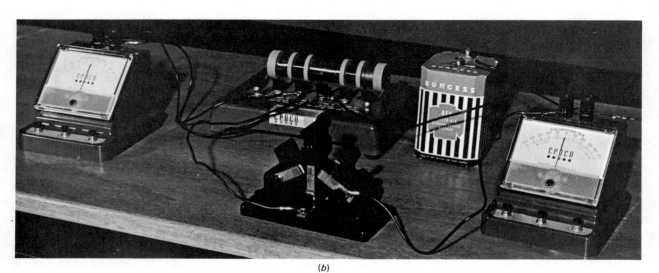

(b)

Figure 42.1. (a) Set of test coils mounted on a board with binding posts for connection into an electrical circuit. *(Central Scientific Co.)* (b) Test coils shown in a circuit for measuring resistance by the voltmeter-ammeter method.

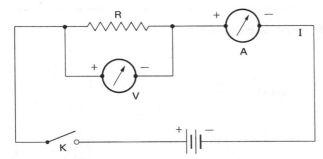

Figure 42.2. Wiring diagram for the voltmeter-ammeter method of resistance measurement.

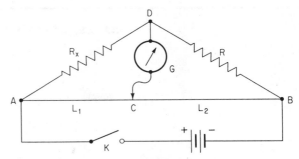

Figure 42.3. Wiring diagram for a Wheatstone bridge circuit for measuring electrical resistance.

Summarize all results in neat tabular form, properly identified.

ANALYSIS AND INTERPRETATION

Discuss what you have learned about methods of measuring and calculating electrical resistance. In what ways could the errors inherent in Part 1 be minimized?

Discuss some important industrial applications of the principles involved in the experiment.

Answer the following questions:

1. State in words the exact meaning of Ohm's law. How were the units of voltage, current, and resistance chosen in order to make Ohm's law true? What are the mks units of the *ampere?* The *volt?* The *ohm?* How are volts and coulombs related to energy?

2. If the diameter of a certain length of a wire is doubled, what happens to its electrical resistance?

3. Calculate the resistance in ohms of a copper wire 10 miles long if its diameter is exactly 0.500 in.

4. How does change in temperature affect electrical resistance? State a mathematical expression for the manner in which the resistivity of a metal varies with temperature, and explain what each term means.

5. Based on some collateral reading, write a paragraph on *superconductivity* (zero resistance), a phenomenon exhibited by some materials at extremely low (*cryogenic*) temperatures.

Figure 42.4. Arrangement of equipment for the measurement of an unknown resistance with a Wheatstone bridge. The unknown resistance coil, a galvanometer, a decade resistance box, and a slide-wire Wheatstone bridge are shown.

NOTES, CALCULATIONS, OR SKETCHES

DATA SHEET

EXPERIMENT 42 ■ MEASURING ELECTRICAL RESISTANCE—OHM'S LAW

Part 1. Voltmeter-Ammeter Method of Resistance Measurement—Set of Coils

Coil description	Current, I amperes (A)	Voltage, V (V)	Diameter		Length	
			mm	mils	m	ft
1.						
2.						
3.						
4.						
5.						

Part 2. Wheatstone Bridge Method of Resistance Measurement—Special Test Coil

Conditions for Null Deflection of Galvanometer

Trial	R (Resistance Box) (Ω)	L_1	L_2	Wire Size		
				AWG gage	Diameter (mm)	Diameter (mils)
1						
2						
3						

NOTE: Since L_1 and L_2 enter into the calculations as a ratio, the units in which they are measured (use same units for both) make no difference.

Test coil resistance as read from ohmmeter_____Ω

DIRECT CURRENT IN SERIES AND PARALLEL CIRCUITS

PURPOSE

To study elementary direct current (dc) electric circuits and the principles of current flow through resistances connected in series and resistances connected in parallel.

APPARATUS

Several resistance coils of unknown resistance (or a board of mounted resistors as supplied by manufacturers of laboratory equipment); Wheatstone bridge, enclosed-dial type (or slide-wire type may be used);[1] dc voltmeter (0 to 3 V); dc ammeter (0 to 1.0 and 0 to 10.0 amp); sensitive galvanometer; digital multimeter, if available; low-voltage dc source (not over 2 V); switch; wire for connections. A typical apparatus setup is illustrated in Fig. 43.1.

INTRODUCTION

Series Circuits If separate items of electrical apparatus (or separate resistors) are connected in such a manner that the source of *emf* forces an electric current through all of them *in sequence,* the circuit is said to be a *series circuit.* Such a circuit is diagrammed in Fig. 43.2. Once steady-state conditions are established in the circuit, the same current exists in all the resistances since there is no alternate path for electron flow.

In a series circuit the total resistance R_T is given by the expression

$$R_T = r_1 + r_2 + r_3 + \cdots + r_n \qquad (43.1)$$

where r_1, r_2, etc., are the resistances of the separate resistors. *In a series circuit the total resistance is merely the sum of the separate resistances.*

Voltage in a Series Circuit Each component of a series circuit (Fig. 43.2) causes a drop in potential (voltage), and the *sum of all the potential drops is equal to the voltage V applied to the total circuit.*

$$V = v_1 + v_2 + v_3 + \cdots + v_n \qquad (43.2)$$

where v_1, v_2, etc., are the potential drops across the separate resistances r_1, r_2, etc. In Fig. 43.2 only v_1 is shown. The voltmeter is connected across r_2, r_3, etc., in turn.

Current in a Series Circuit Current in a series circuit is the same at all points in the circuit, since no alternative

path for electron flow is offered by such a circuit. *The total current from and to the source is the same as that through any one of the circuit components.* As an equation

$$I = i_1 = i_2 = i_3 = \cdots = i_n \qquad (43.3)$$

Parallel Circuits When the components of an electric circuit are connected in such a manner that alternative paths for electron flow are afforded, the circuit is said to be a *parallel circuit.* Such a circuit is diagrammed in Fig. 43.3. In parallel circuits the current paths are not sequential and electrons "have a choice" among two or more paths. Like human beings most of them will "take the path of least resistance." Since several paths are available for the current (r_1, r_2, and r_3 in Fig. 43.3) instead of just one, the *total resistance of a parallel circuit is less than the resistance of any one of the alternative paths.* The total resistance of a parallel circuit is given by

$$\frac{1}{R_T} = \frac{1}{r_1} + \frac{1}{r_2} + \frac{1}{r_3} + \cdots + \frac{1}{r_n} \qquad (43.4)$$

In words, *the reciprocal of the total resistance of a parallel circuit is equal to the sum of the reciprocals of the separate resistances.*

Voltage in a Parallel Circuit Since voltage is a measure of electrical pressure, and since at every point in a parallel circuit the electron current "has a choice" of path, the effect is that the *electrical pressure (voltage) in a parallel circuit is the same across any component of the circuit as it is across the circuit as a whole.*

$$V = v_1 = v_2 = v_3 = \cdots = v_n \qquad (43.5)$$

Current in a Parallel Circuit The electron current leaving the source divides and pursues various paths through the separate resistances of a parallel circuit, the current in each being inversely proportional to the resistance of that component. All paths unite again, however, to complete the circuit to the source. The *total current in a parallel circuit is the sum of the currents in the separate components.*

$$I = i_1 + i_2 + i_3 + \cdots + i_n \qquad (43.6)$$

One purpose of the investigations to follow is to verify each of the expressions [Eqs. (43.1) through (43.6)] above.

---

[1]A digital multimeter may be used, if available.

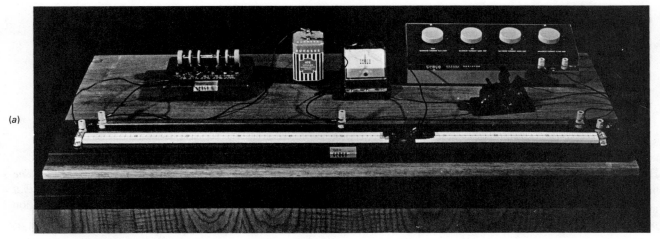

Figure 43.1. Slide-wire Wheatstone bridge and associated equipment assembled to measure the resistance of one of the coils of wire on a test board. The decade resistance box at the right serves to supply the known resistance. (b) Self-contained "Wheatstone bridge in a box." (Central Scientific Co.)

MEASUREMENTS

Part 1. Resistance Measurements

With the Wheatstone bridge (two types are illustrated in Fig. 43.1) determine the resistance of each of the separate resistance coils to an accuracy of three significant figures. Use an emf not greater than about 1.5 V. (The use of the slide-wire Wheatstone bridge was explained in Experiment 42.)

Part 2. Series Circuit Measurements

Connect all the resistors in series. Use the shortest possible length of connecting wire, and make sure that it is large-size low-resistance wire.

Measure the total resistance of the circuit with the Wheatstone bridge. Check this measurement with a digital multimeter.

With the voltmeter and ammeter take all the readings required to verify Eqs. (43.2) and (43.3). Be sure to use meter ranges which allow *accurate* readings of voltage and current. Use an emf (voltage) which will give a significant current flow in each case, but not above 2 A in any case. (A *voltmeter* is always connected in parallel, or *across* the component whose potential drop is to be mea-

sured. An *ammeter* must be connected in series, so that the current which is to be measured will flow *through* the meter.)

Part 3. Parallel Circuit Measurements

Note: For conclusive results in this part of the experiment, it is best not to use small resistances (less than 1 Ω) in the same circuit with large resistances (more than 50 Ω).

Connect all the resistors in parallel. Measure the total resistance of the parallel circuit with the Wheatstone bridge. Use an emf of not more than 1 V. Check total resistance with a digital multimeter.

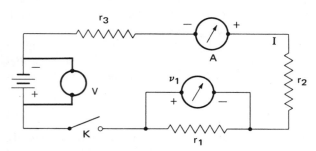

Figure 43.2. Diagram of a typical (and simple) series circuit, with ammeter and voltmeter connections indicated.

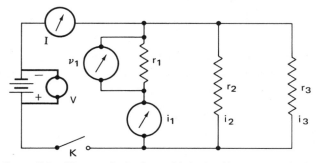

Figure 43.3. Diagram of a simple parallel circuit, with ammeter and voltmeter connections indicated.

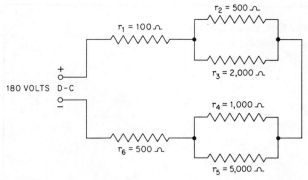

Figure 43.4. A very simple series-parallel (combination) circuit.

With the voltmeter and the ammeter take all the readings required to verify Eqs. (43.5) and (43.6). Again, take care in the selection of the meter ranges, so that accurate readings will be obtained. Observe reasonable limitations on current as before.

CALCULATIONS AND RESULTS

From the measurements in Parts 1 and 2, perform all the calculations necessary to check the applicability of Eqs. (43.1) through (43.3) to series circuits. Compute the percent error of your experimental results from the results predicted by the equations.

From the measurements in Parts 1 and 3, check the applicability of Eqs. (43.4) to (43.6) to parallel circuits, and compute the percent error from expected results.

Summarize all findings in such a way that they can be readily interpreted by anyone reading your report.

ANALYSIS AND INTERPRETATION

What have you learned about series and parallel circuits from this experiment? Summarize carefully.

Discuss the sources of error in the experiment. Why were the limitations on current flow suggested?

What effect (if any) does the presence of the ammeter and the voltmeter in the circuit have on the measurements of current and voltage drop? Discuss this matter in some detail.

What kinds of electric machinery or electric apparatus are ordinarily connected in series? In parallel?

Frequently, it is necessary to connect a component of electronics equipment in a combination (series and parallel) circuit. Calculate the total resistance, the total current, and the current flowing in each component of the circuit in Fig. 43.4.

DATA SHEET

EXPERIMENT 43 ■ DIRECT CURRENT IN SERIES AND PARALLEL CIRCUITS

Part 1. Measuring the Separate Resistances

Coil number and description	Resistance, ohms, Ω	
	Wheatstone bridge	Digital multimeter
1.		
2.		
3.		
4.		
5.		
6.		

Part 2. Series Circuit Measurements

Resistance of entire circuit_____Ω

(Student prepares own data table.)

Part 3. Parallel Circuit Measurements

Resistance of entire circuit_____Ω

(Student prepares own data table.)

EXPERIMENT 44

ELECTRICITY BY CHEMICAL ACTION

PURPOSE

To study the construction and action of the simple voltaic cell and the lead storage cell.

INTRODUCTION

In order that a current of electricity will flow continuously along a wire, a steady supply of electron charges must be available at one end and a condition of deficiency of electron charge (positive charge) must exist at the other end. The difference in charge of the ends of the wire is called *potential difference,* and it is measured in *volts*. The rate of flow of electric charge is called *current,* and it is measured in *amperes*. A current of one ampere in a wire requires the flow of 6.24×10^{18} electrons per second past any point in the wire. The *resistance* to electron flow exhibited by a wire is measured in *ohms*. Voltage, current, and resistance are the major circuit factors in direct-current circuits, and they are related by Ohm's law, which was studied in Experiment 42.

There are two common methods by which a continuous supply of electric charges and a potential difference are produced: one method makes use of chemical changes, and the other consists in converting mechanical energy into electric energy by rotating a coil of wire in a magnetic field. The first method takes place in an electric cell or battery, while the second method takes place in an electric generator. The first method will be studied in this experiment. (*Caution:* Be extremely careful in handling the acid electrolyte used in the experiment! It is highly corrosive).

Part 1.   The Voltaic Cell

Apparatus   Battery jar with clamps; strips of copper and zinc; dc voltmeter (0 to 3 V); dilute sulfuric acid (about 1 part concentrated H_2SO_4 to 15 parts water); switch; flashlight lamp (1.5 V) with a socket and binding posts; wire for connections; sandpaper.

Theory   Any metal which is "attacked" by an acid will surrender to the electrolyte positive (+) ions of the metal. Thus, when a copper rod is dipped into a dilute acid, copper ions (positive) go into solution in the electrolyte and the electrons surrendered by the copper atoms which become ionized are left on the copper rod. The rod itself thus acquires a negative charge, and the electrolyte acquires a positive charge (see Fig. 44.1).

If some different metal (e.g., zinc) is placed in the same solution, the same process of ionization occurs, but

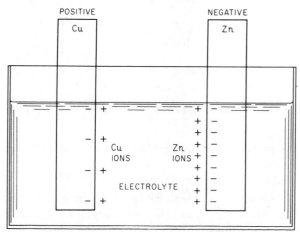

Figure 44.1.   Two unlike metal strips in an electrolyte result in chemical changes that produce free electrons. A simple voltaic cell is the result.

acid attacks zinc much more actively than it does copper, and many more zinc ions than copper ions go into solution. A zinc rod will therefore become *more negative* than a copper rod when the two are placed in the same acid solution (electrolyte). Thus, though both copper and zinc are negative with respect to the electrolyte, the copper is less negative and is therefore *positive* with respect to the zinc. Since a potential difference exists between the two metals (called *electrodes*), there will be an *electron current* from the zinc to the copper if the two electrodes are connected by a conductor. [The *conventional* current (+ to −) would be from copper to zinc.] As electrons leave the zinc metal, more zinc atoms go into solution, thus freeing more electrons so that a steady current can be maintained. The process will continue as long as the zinc metal remains and as long as the strength of the acid is maintained.

Such a cell is called a *voltaic* cell, in honor of Alessandro Volta, who first studied them about the year 1800. Cells of this type are also called *primary cells*.

Method and Observations   Fill the battery jar about three-fourths full of the dilute (15:1) acid. Polish the copper (*anode*) and the zinc (*cathode*) strips with fine sandpaper, and insert them in the clamps. Lower them into the electrolyte, and connect the voltmeter to the binding posts (Fig. 44.2).

What is the emf of the cell, as read on the voltmeter?

Leaving the voltmeter across the cell, connect the

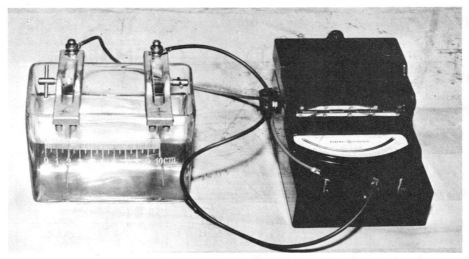

Figure 44.2. Battery jar, electrolyte, and electrodes set up as a primary (copper-zinc) voltaic cell.

flashlight lamp socket and lamp to the cell with a switch in the circuit.

Make the following observations:

1. Which strip reacts with the acid more on open circuit? Which on closed circuit? Note carefully the evidences of chemical reactions on both metal surfaces.

2. What voltage is obtained when the switch is first closed? Does it stay constant as the flashlight lamp glows?

3. Note the effect on the emf of removing the electrodes momentarily and replacing them.

4. What is *local action?* Explain fully what causes it, and the effect it has on the emf of the cell.

5. What collects on the copper strip (anode) after the circuit has been closed a few seconds? Explain what is meant by *polarization* of a cell. Why does this effect cause a drop in *emf?*

6. Correct for local action by "amalgamating" a zinc strip with mercury. (*Be careful with mercury!* It is poison. Wash your hands after completing this part of the experiment). Reassemble the cell, and note the difference, if any, in constancy of the emf.

7. Explain the difference between the value of the emf as first measured and its value while current is flowing through the lamp.

Part 2.   The Lead Storage Cell

Apparatus   Battery jar with clamps; two lead strips for electrodes; dc voltmeter (0 to 3 V); dc ammeter (0 to 1 and 0 to 10 A); buzzer or doorbell; switch; sandpaper; dilute sulfuric acid (about 8 parts water to 1 part concentrated H_2SO_4); laboratory dc power supply (6 V); wire for connections.

Theory   Cells such as the voltaic cell studied in Part 1 are called *primary cells* since they furnish an electron current only as their materials are used up through chemical changes. In this second investigation, we will study a type of cell in which the chemical changes are *reversible*. A certain chemical change occurs as the cell furnishes elec-

trons, and this chemical action is reversed as electron charges are *furnished to the cell* from an external source. Such a cell, although its materials are "used up" as it is *discharging*, replenishes itself chemically as it is *charged*. Such cells are called *storage cells*, or *secondary cells*.

The most common kind of storage cell is the *lead cell*, in which two electrodes or *plates* (or groups of plates) are immersed in dilute sulfuric acid. The active material of the cathode is spongy metallic lead, and that of the anode is lead peroxide (PbO_2).

As the cell is discharged, both plates undergo chemical change and become coated with lead sulfate. Upon being charged (electrons supplied from an external source), the cathode reverts to spongy lead, and the anode to lead peroxide. Upon discharge, the electrolyte becomes more dilute, and its specific gravity drops nearer 1.00. When being charged, the sulfate ion recombines with hydrogen to form sulfuric acid and the specific gravity of the electrolyte increases. At full charge, the specific gravity of the electrolyte in an automobile-type lead cell is about 1.3. Such a lead cell has an emf of about 2.12 V. Single cells are connected in series to provide the 12-volt emf required for auto electrical systems.

Method and Observations   Clean the lead strips thoroughly with fine sandpaper (to remove oxides and other impurities), and set up the cell, with the lead strips immersed in the 8:1 sulfuric acid electrolyte.

Connect it to a voltmeter. Is there any evidence of current flow or chemical reaction?

After a few minutes lift the lead strips out of the electrolyte, and examine them. Has any visible change occurred?

Charge the cell from the laboratory dc source (6 V) for about 2 min. The charging circuit should contain an ammeter and a voltmeter. Note which electrode is positive. Make the following observations while the cell is being charged.

1. Record the charging voltage.

2. Record the charging current, in amperes.

3. At which electrode is chemical activity more evident? What is the gas being given off at each electrode?

4. Calculate the internal resistance of the cell. Test the effect on the charging rate (and therefore on the cell's internal resistance) of varying the distance between the electrodes.

After charging is stopped, examine the lead strips carefully. What change in color is noted on the anode? The cathode? Measure the emf of the cell itself.

Now short circuit the cell for about 5 min or until it shows near-zero emf. What changes have occurred on the plates during discharge?

Charge the cell again for about 10 min. After disconnecting the charging circuit, measure the emf of the cell again, and again inspect the electrodes carefully, noting the color and texture of the coating on each. On which is the lead peroxide found?

Connect the buzzer, the ammeter, and the voltmeter to the cell. Close the switch, and note the changing values of current and emf as the cell runs down while ringing the buzzer.

Return the electrolyte to the storage jar, rinse all parts of the cell thoroughly, dry them, and return to the equipment room.

5. Examine thoroughly and make sketches of a standard 12-V auto battery (cutaway, if available). How many plates does it have?

DATA SHEET

Plan your own Data Sheet for this experiment.

PRESENTATION OF RESULTS

Answer all of the questions expressed and implied in the Observations sections of both Parts 1 and 2.

Make sketches of each cell, labeling everything.

For each cell, using further sketches as necessary, explain thoroughly the processes by which chemical energy is transformed into electric energy. For the lead storage cell, explain how the process is reversible. Separate sketches for the charging situation and for the discharging situation will be required.

ANALYSIS AND INTERPRETATION

1. Summarize what you have learned about producing electricity from chemical action. In your discussion be sure to touch on the following:

a. The nature of an *ion*.

b. The difference between *primary* cells and *secondary* cells.

c. How is it possible to obtain electron flow from the copper-zinc cell, when both electrodes have an excess of (negative) electrons?

d. Since the processes involved in the lead storage cell are reversible, why do such cells "wear out" in actual practice?

e. How is it possible to tell (roughly) the condition of charge of a lead storage cell with a hydrometer?

2. Give a complete description, with sketches, of a real automobile-type lead storage battery. Why are there so many plates of large area? What are the separators for? What electrolyte is used? What should be its specific gravity at full charge? A storage battery is usually rated in *ampere-hour* capacity. What does this mean?

3. Auto manufacturers are producing test models of electric automobiles in order to reduce air pollution. What are some of the major advantages and disadvantages of lead-cell-operated electric autos?

4. Sometimes "dead" batteries of autos are energized via "jumper cables" from a good battery, to start the engine. If a spark occurs near the battery an explosion could result. Why? What precautions should be taken? Explain clearly how the "jumper cables" should be connected.

HEATING EFFECT OF ELECTRIC CURRENT—JOULE'S CONSTANT

PURPOSE

To study the heating effect of an electric current, and to determine the heat equivalent of electrical watts as a check on the value of Joule's constant J.

APPARATUS

Calorimeter; electric resistance heating element (immersion heater); stop clock; rheostat; dc ammeter (0 to 1, 1 to 10 amp); dc voltmeter (0 to 12 V); thermometer (accuracy to 0.2°C); ice; switch; source of steady dc voltage (12 V or more); wire for connections.[1]

INTRODUCTION

The student should first review the related phenomena of heat exchange and calorimetry, and recall that 1 kcal is the amount of heat required to raise the temperature of 1 kg of water 1 C°.

Both mechanical power and electric power are expressed in *watts*. Mechanically,

$$1 \ \frac{\text{joule}}{\text{second}} = 1 \text{ watt}$$

Electrically,

1 volt × 1 ampere = 1 watt

or

Power (watts) = volts × amperes

or

$$P = VI \qquad (45.1)$$

But *power* is the time rate of doing *work* or of expending *energy*. Consequently *electric energy* is given by

$$W = \text{power} \times \text{time} = VIt \qquad (45.2)$$

where

W = work (energy), in joules (J)
I = current, amp
V = volts
t = time, seconds

---

[1] (a) The experiment can be done with ac if desired. In this case, an autotransformer and ac meters would be used. (b) If the apparatus illustrated in Figs. 45.1 and 45.2 is not available, satisfactory results can be obtained using a styrofoam cup for the calorimeter (provide insulating cover locally) and an immersion heater of the type used to heat a cup of water for making instant coffee or tea. These are available at many drug stores and department stores.

It will be recalled (see your physics text) that the basic relationship between heat and work is expressed by Joule's equation

$$W = JH \qquad (45.3)$$

where

W = work, joules or ft · lb
H = heat, kcal or Btu
J = Joule's constant, (the mechanical equivalent of heat)
  = 4186 J/kcal
  = 778 ft lb/Btu

Note use of J (italic) for Joule's constant, and J for the energy unit *joule*.
Substituting from Eq. (45.3) in Eq. (45.2)

$$JH = IVt$$

or

$$H_{(\text{kcal})} = \frac{IVt}{4.186 \times 10^3}$$

or

$$H_{(\text{kcal})} = 2.39 \times 10^{-4} \, IVt \qquad (45.4)$$

In the experiment, an electric resistance heating element is immersed in a known mass of water in a calorimeter. Assuming that all the heat generated in the resistance heating coil is transferred to the water, and assuming no heat losses from or gains to the calorimeter, then

$$H = (m_w c_w + m_{\text{cal}} c_{\text{cal}} + WE_{hc})(t_f - t_c) \qquad (45.5)$$

where

H = heat, kcal
m_w = mass of water, kg
c_w = sp ht of water = 1.00 kcal/(kg · C°)
m_{cal} = mass of inner calorimeter cup, kg
c_{cal} = sp ht of calorimeter material
WE_{hc} = water equivalent of the electric resistance heating element, kg (available from supplier)
t_c = temperature of cold water and calorimeter, °C.
t_f = temperature of water and calorimeter, after heating, °C.

MEASUREMENTS

1. Weigh the empty calorimeter cup (see Fig. 45.1). Then fill it about two-thirds full of cold tap water. If the tap water is not several degrees colder than room temperature,

---

Figure 45.1. Right to left: inner and outer calorimeter cups, insulating ring support for inner cup, and electric resistance immersion heater with stirrer. *(MSOE Academic Media Services)*

chill it with chipped ice to about 10°C below room temperature. Be sure all ice has melted before beginning a test run. Weigh the calorimeter cup with the water in it to determine the exact mass of cool water.

2. Hook up the heating circuit as shown in the photograph in Fig. 45.2 and in the circuit diagram of Fig. 45.3. Do not close the switch.

3. Assemble the calorimeter, and immerse the heating element in the water. Insert the thermometer in the water, and after a short interval of stirring, read the initial temperature t_c accurate to 0.2°C. Note the room temperature.

4. Close the switch, start the timer, and for a measured time interval, pass current through the heating element,

stirring frequently. Heat until final temperature is as much above room temperature as the initial temperature was below room temperature. Keep the current constant throughout the trial by means of the control rheostat. Open the switch, and stop the clock simultaneously. Stir and record the highest reading of the thermometer, again to the nearest 0.2°C.

5. Take three more trials, using different initial amounts of water, at different initial temperatures (using ice as required to obtain colder water), and changing the current for each trial by different rheostat settings.

Record all data on the Data Sheet.

CALCULATIONS

1. From Eq. (45.2) calculate the *electric energy* dissipated in each trial, in joules.

2. From Eq. (45.5) calculate the heat absorbed by the water in each trial, in kilocalories.

3. Then, from Joule's equation [Eq. (45.3)], determine an experimental value of J for each trial, in joules per kilocalorie.

4. Average the results of the four trials, and compare your value of J with the accepted value (see Appendix) by calculating the percent error.

5. As 1 W of electric power is dissipated in a resistance heater, how many Btu per hour of heat are produced? Compare your answer with the value given in Table VII of the Appendix.

ANALYSIS AND INTERPRETATION

1. Discuss the principal sources of error, and attempt to justify the discrepancy (if any) between your result and the accepted value. Explain why it is important to start the

Figure 45.2. Electric resistance heating element, thermometer, and stirrer ready for immersion in the water in the calorimeter, just prior to a test run. Note that the ammeter and the resistance heater are connected in series. The voltmeter (not shown) is connected *across* the binding posts of the heating coil.

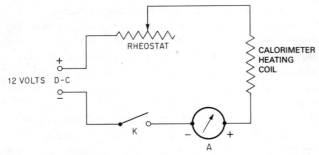

+
12 VOLTS D-C
−

RHEOSTAT

CALORIMETER
HEATING
COIL

K

− A +

Figure 45.3. Circuit diagram for making electrical connections to the calorimeter heating element.

heating process with water several degrees colder than room temperature.

2. Explain clearly the difference between electric *power* and electric *energy*.

3. Prove, by dimensional analysis, that 4186 J/kcal is the equivalent of 778 ft · lb/Btu.

4. Based on electric-resistance heating, what would the electric bill be for heating 500 gal of water from 40°F to 212°F if electric energy costs 8.5 cents per kilowatt-hour (kwh)? Assume 100 percent efficiency.

DATA SHEET

EXPERIMENT 45 ■ HEATING EFFECT OF ELECTRIC CURRENT—JOULE'S CONSTANT

Trial	Current, I (amp)	Voltage, V (volts)	Mass of water, m_w (kg)	Heating time, t (seconds)	Initial temperature, t_c (°C)	Final temperature, t_f (°C)
1						
2						
3						
4						

Mass of calorimeter cup m_{cal} _____ kg

Specific heat of calorimeter material c_{cal} _____ kcal/kg · C°

Water equivalent of heating element WE_{hc} _____ kg (provided by the supplier or by your instructor)

Room temperature _____ °C

NOTES, CALCULATIONS, OR SKETCHES

EXPERIMENT 46

STUDIES OF ELECTROMAGNETIC INDUCTION

PURPOSE

To study the laws of electromagnetic induction, Lenz's law, and the influence of iron on a magnetic circuit.

APPARATUS

Sensitive galvanometer; decade resistance box; Gilley induction study apparatus; rheostat (100-Ω); strong (Alnico) bar magnet; small compasses; key or switch; several feet of cotton-covered bell wire; source of laboratory direct current.

INTRODUCTION

Oersted's discovery of the relationship between electric currents and magnetic fields (about 1820) opened a vast new era of scientific advance. The fact that an electric current in a wire sets up a magnetic field surrounding the wire has led to many of the scientific developments of the past 150 years.

The converse effect—electricity from magnetism—was discovered about 11 years later (1831) by Faraday. The discovery that relative motion between a conductor and a magnetic field will produce an electric current in the conductor made possible the production of electric energy directly from mechanical work.

An electric current can be produced in a conductor by relative motion between the magnetic field and the conductor, or by any process which causes a change in the magnetic flux (lines of force) passing through or "cutting" the conductor (see Fig. 46.1). A magnet may be moved relative to a coil of wire, or the coil may be moved relative to the magnet. Or, the magnetic field may be produced by current in one coil, and the flux produced may "cut" the conductors of a second coil. It should be emphasized, however, that there must be relative motion between the magnetic field and the conductor. It is only dur-

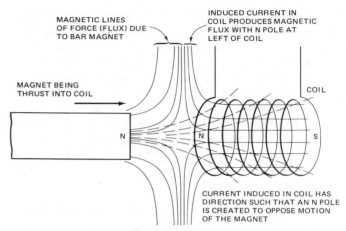

Figure 46.2. Diagram illustrating Lenz's law.

ing the interval when current is *changing* in the first coil (thus, a magnetic field will be building or collapsing) that an emf will be induced in a neighboring coil.

A current which is induced in a secondary coil will then immediately set up a magnetic field of its own. This magnetic field will have a polarity which opposes any increase of the original (inducing) field (i.e., N to N, or S to S). Consequently, it is necessary to do mechanical work to overcome the force of repulsion between the increasing original field and the field of the secondary (see Fig. 46.2 and the discussion of Lenz's law, below).

In all phenomena of magnetism the effects are multiplied by the presence of iron "cores" in the coils of wire. Iron makes the magnetic flux density greater and it therefore increases the effects of electromagnetic induction.

The fact that electric energy cannot be produced without the expenditure of some other form of energy (in this case, mechanical energy) is consistent with the general law of conservation of energy. Specifically, if a loop or coil of wire is placed in a region of magnetic flux, and *if the magnetic flux is changing with time,* an emf will be produced in the coil. The magnitude of the emf will be directly proportional to the time rate of change of magnetic flux. *If the flux is constant, no emf will be produced.* The induced emf is given by the expression

$$V = -N\frac{\Delta\Phi}{\Delta t} \tag{46.1}$$

where

V = average value of the induced emf, V

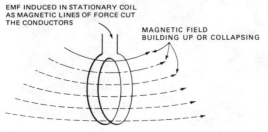

Figure 46.1. A current is induced in a coil of wire if it is "cut" by the lines of force of a magnetic field as they are moved, or as they are established or collapsed.

N = number of turns on the coil of wire
$\Delta\Phi$ = change of flux, webers (Wb)
Δt = time for the change in flux to take place, seconds

If the magnetic flux is increasing with time, the emf induced will be in one direction; if it is decreasing with time, the emf will be in the other direction. Lenz's law, a statement of the law of conservation of energy as it applies to electromagnetic induction, states that *electromagnetically induced currents will always have a direction such that the magnetic field set up by the induced currents will oppose the action that produces them.* (See Fig. 46.2.) Lenz's law is, then, merely a statement of the law of conservation of energy as applied to the mechanical-to-electric or the electric-to-mechanical energy conversion process. The minus sign in Eq. (46.1) indicates this opposition.

The observations to follow are largely qualitative. Take down your observations in detail, and draw sketches to illustrate the effects you observe.

MEASUREMENTS

Part 1. Direction of Galvanometer Deflection for a Known Current Direction

Set up a potential divider circuit as shown in Fig. 46.3. Use the laboratory dc outlet as a source of low voltage (about 1 V). The rheostat should be of less than 100 Ω, and the slide should be set to tap off just one or two turns. The setting on the decade resistance box should be quite high at first. Reduce it as necessary to obtain a reasonable galvanometer deflection. Close the key, and note the direction of galvanometer deflection and the direction of current (+ to −). Mark which is the + direction for the galvanometer and use this throughout the experiment.

Part 2. Induction in a Loop of Wire

Make a single loop of the bell wire about 1 in. in diameter, and connect it to the galvanometer. Thrust a strong bar magnet through it quickly, noting polarity of the magnet and direction of flow of induced emf, as indicated by the

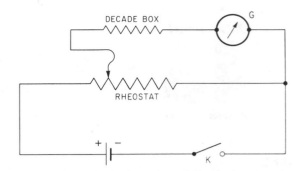

Figure 46.3. Circuit diagram for a voltage divider circuit to determine the direction of galvanometer deflection for a known current direction.

galvanometer. Now withdraw the magnet from the loop sharply, noting the direction of current flow in the loop. By carefully drawn diagrams, show how the direction of the current in the loop is consistent with Lenz's law.

What happens if there is no relative motion of magnet and loop?

Now make coils of the same wire containing 2, 5, and 10 turns, and repeat the observations with each. How is the emf produced related to the number of turns?

Part 3. Studies with the Gilley Apparatus

See Fig. 46.4.

a. Connect one Gilley coil in series with a key, a rheostat, and the dc source in such a way that the current direction (+ to −) in the coil is clockwise as viewed from the side opposite the binding posts. Check the polarity of this electromagnet with the small compass. Does it conform to the "right-hand rule"?

Connect the other Gilley coil to the galvanometer terminals through the decade box so that the sensitivity of the meter may be varied.

Now set the two coils back to back. Set the rheostat for minimum effect and the decade box at maximum. Close the switch, and note galvanometer deflection. Decrease the resistance box setting as necessary to get a good (say half-scale) galvanometer deflection each time the switch is opened or closed.

Figure 46.4. Arrangement of equipment for induction studies with the Gilley coils. The rheostat controls current to the primary coil, and the decade resistance box setting protects the galvanometer in the secondary coil circuit. The U-shaped iron core is shown in place completing the magnetic circuit between the Gilley coils. This arrangement of apparatus relates to Part 3*d*, under measurements.

Is the induced current (secondary coil) in the same or opposite direction as that in the (primary) coil connected to the dc source? Explain in terms of Lenz's law. Use diagrams. Note the galvanometer reading as the switch is closed and then as the switch is opened. Explain both. Repeat with coils separated about 1 in. Explain.

b. Set the coils up as in *3a,* and close the switch. When the galvanometer needle has come to rest, move the rheostat slider quickly to maximum resistance. Explain what happens. Now move slider back to minimum resistance, and note the result. Explain what you observe.

c. Set the coils as in *3a,* and decrease the galvanometer sensitivity by increasing the resistance box setting until a kick of just one or two scale divisions is obtained as the switch is closed.

Now insert the soft iron core through both coils, and close the switch again, noting the deflection. Explain. Now pull the coils about 2 in. apart, but with the iron core still in. Close the switch and note deflection. Does the distance between the coils when an iron core is present appreciably affect the reading? Explain. Check the polarity of the secondary coil with the small compass during the buildup of the induced emf. Does it conform to Lenz's law?

d. Set the two coils side by side ready to receive the U-shaped iron core. Before inserting the core, close the switch and note the deflection. Adjust rheostat and resistance box as necessary to get a deflection of two or three scale divisions. Now place the U-core in the coils. Close the switch and note the deflection. What is proved with regard to the path taken by magnetic flux? How does iron complete a magnetic circuit?

PRESENTATION OF RESULTS

There is no formal Data Sheet.

Describe by carefully drawn sketches and *briefly* in words every step of the experiment as outlined above. State what you observed, and explain *why* each effect happened as it did. There are no numerical calculations. These sketches and explanations constitute your findings, and your report. Draw them neatly; label everything.

ANALYSIS AND INTERPRETATION

Explain in considerable detail what you have learned about electromagnetic induction. With the aid of sketches show that you understand Lenz's law.

After some collateral reading, explain what is meant by the terms *self-induction, mutual induction, back emf.*

Why is a constant current in a primary coil unsuitable for producing a usable output from a secondary coil?

Answer the following:

1. As a bar magnet is inserted into a coil of wire an electric current is induced in the wire. This current will provide *electric energy.* What is the source of that energy?

2. Why is it that an electric generator may be rotated easily on open circuit but becomes very hard to turn when it is connected to a load? Think this out very carefully!

3. A conducting loop of 50 turns of wire is located in a magnetic field where the flux changes from 6×10^{-4} Wb to 3×10^{-5} Wb in 0.012 s. What is the average induced *emf?*

4. In energy transformations of all kinds, there are losses—heat cannot be completely converted to work and mechanical energy cannot be completely converted to electric energy. In what form (of energy) does the loss appear when mechanical energy is converted to electric energy?

5. In an automobile ignition system, electromagnetic induction is used to obtain a very high voltage for the spark plugs from the 12 V dc available at the terminals of the auto's storage battery. Discuss how this is accomplished. Include a labeled diagram that shows the battery, the ignition coil, and other important components.

NOTES, CALCULATIONS, OR SKETCHES

MAGNETIC INDUCTION IN AN IRON ROD—HYSTERESIS

PURPOSE

To study the magnetization behavior of soft iron and to plot the hysteresis curve for a soft iron cylindrical rod, acting as the core of a long solenoid.

APPARATUS

Hysteresis apparatus, consisting of a long solenoid with primary and secondary coils, mounted on a base with four resistances and short-circuiting switches; a 6-V dc source; dc ammeter; rheostat; ballistic galvanometer; resistance box; ac (120-V) source; low-voltage (12-V) transformer; soft-iron test rod; micrometer calipers; wire for connections; switches.

INTRODUCTION

First, an elementary review of electrically-induced magnetism. The following paragraphs deal with the topic rather superficially, but they will serve as a guide for you as you refresh your memory from your textbook and your lecture notes.

The Terminology of Magnetic Induction

An elementary study of magnetic fields was carried out in Experiment 41. Magnetic "lines of force" were plotted from observations on the behavior of iron filings near magnets, and by testing for a magnetic field with small compasses.

Magnetic Field Strength. When direct current flows in a long solenoid (i.e., the length is great compared to the diameter), a magnetic field is produced within and around the coil (see Fig. 47.1). If a current of I amperes flows through a uniformly wound, hollow-core solenoid of length L, with N turns, the strength of the magnetic field along the center line, or axis, of the solenoid may be calculated from

$$\mathbf{H} = \frac{NI}{L} \qquad (47.1)$$

where H is *field intensity* (magnetizing force) in ampere-turns/meter, I is in amperes, and L in meters.

The *strength of a magnetic field* is related to the number of lines of force in the direction N to S of the magnetic field. Another term for "strength of a magnetic field" is *field intensity*. Magnetic field strength has a definite direction (N to S) and is therefore a vector quantity, indicated by **H**.

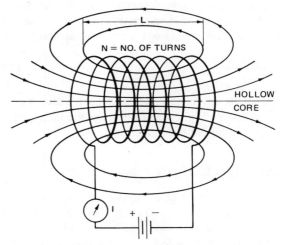

Figure 47.1. Schematic diagram of a magnetic field in and around a current-carrying wire wound as a long solenoid. with an air core, $\mathbf{H} = \frac{(NI)}{L}$; With an iron core of permeability μ, $\mathbf{B} = \mu \frac{(NI)}{L}$

Magnetic flux The strength of a magnetic field within a specific *material* (not air) is ordinarily expressed by the term *magnetic flux*, which is a term describing the *total* quantity of lines of force in the magnetic field established by the magnet. The symbol for magnetic flux is Φ, (Greek *phi*) and its unit (mks) is the *weber* (Wb). A weber is dimensionally equivalent to 1 newton-meter per ampere (1 Wb = 1 N · m/amp).

Magnetic flux density The extent to which magnetic lines of force are *concentrated* in a given area of a (nonair) material is described by the term *magnetic flux density*, the symbol for which is **B**. It is also a vector quantity. It is a measure of the number of lines of force *per unit area*, within a region of the material, cutting a plane which is perpendicular to the direction of the magnetic field, N to S.

It follows from these definitions that magnetic flux density can be expressed as

$$\mathbf{B} = \frac{\Phi}{A} \qquad (47.2)$$

The unit of magnetic flux density is the *weber per square meter* (Wb/m²). The Wb/m² is also called the *tesla* (T). An older (cgs) unit of magnetic flux density is the *gauss* (1 gauss = 10^{-4} Wb/m²). 1 T = 10^4 gauss = 1 Wb/m².

Since one newton-meter (N · m) is equal to one *joule*,

and one ampere is equal to one *coulomb per second*, it follows that

$$1 \frac{Wb}{s} = 1 \frac{N \cdot m}{amp \cdot s} = 1 \frac{joule}{\dfrac{coulomb}{s} \cdot s} \qquad (47.3)$$

$$= 1 \frac{joule}{coulomb} = 1 \text{ volt}$$

The weber, then, represents the magnetic flux that must be cut by a conductor in one second to produce one volt of induced emf, or 1 Wb/s = 1 volt. Consequently,

$$V = -\frac{\Delta\Phi}{\Delta t} \qquad (47.4)$$

The minus sign merely indicates that the law of conservation of energy applies in electromagnetic induction, just as it does in all energy transformations.

Electromagnetism—The Solenoid

A solenoid is a coil of many turns of current-carrying wire wound around a nonconducting form of rigid paper, wood, or ceramic, as indicated in Fig. 47.1. The space down the axis of the solenoid may be hollow (air core) or it may be filled by another material, often iron. As electron flow occurs in each turn of wire, a magnetic field is established around that turn (or loop) of wire. Each of these unitary magnetic fields is additive (vectorially) with the fields set up by other current-carrying loops in the coil, and the result is a more-or-less uniform magnetic field down the length of the solenoid, parallel to the axis of the core. For a *long* solenoid, (length L being 8 to 10 times the diameter) the following expression for the magnetic flux density applies. It is given here without derivation:

$$\mathbf{B} = \frac{\mu NI}{L} \qquad (47.5)$$

where N is the number of turns, I is the current (amperes), L is the length (meters), and μ is the *magnetic permeability* of the core material. The fraction $\dfrac{NI}{L}$ will be recognized as $\mathbf{H}$ in Eq. (47.1).

Magnetic permeability is a measure of how readily a material or substance can be magnetized, that is, how susceptible it is to having a strong magnetic flux induced in it by the presence of a nearby or surrounding magnetic field. In this case, we are concerned with the core of a solenoid and the magnetic field produced in it by the current-carrying coil that surrounds it.

The magnetic permeability of materials varies widely, free space having the basic value, designated by μ_0.

μ_0 (free space) $= 4\pi \times 10^{-7}$ Wb/A $\cdot$ m.

For air, $\mu_{air} \cong \mu_0$. On a *relative* basis, μ_0 being assigned a value of 1.0000, $\mu_{air} \cong 1.0$.

Iron has a very high permeability, and replacing an air

core with an iron core in a solenoid increases the magnetic flux $\mathbf{B}$ dramatically.

The *magnetic field strength* $\mathbf{H}$ created by a current-carrying coil is related to the *magnetic flux density* $\mathbf{B}$ produced in the core of that coil by the equation

$$\mathbf{B} = \mu\mathbf{H} \qquad (47.6)$$

where μ is the permeability of the core material. In an iron core, the flux density may be thousands of times greater than the field intensity would be if the core were hollow (air).

Equation (47.6) would seem to indicate that there is a linear relationship between $\mathbf{B}$ and $\mathbf{H}$, but this is not the case, because the value of μ for any core material changes as $\mathbf{H}$ changes. The *magnetic field intensity* $\mathbf{H}$ can be thought of as being due to the *current* in the solenoid coil; and the magnetic flux density $\mathbf{B}$ as being the *total* field in the core, after the increase in flux due to the *material* of the core (iron, for this experiment) is added.

The *total flux* Φ (webers) through a solenoid with an air core can be derived [from Eqs. (47.2) and (47.5)] as

$$\Phi = \mu_{air}\frac{NI}{L} A \qquad (47.7)$$

where $\mu_{air} \cong 1.0$.

The total flux Φ in a solenoid with an iron core is the sum of two fluxes—that through the iron and that through the remaining air space in the core—and is given by

$$\Phi = \mathbf{B}\, a + \mu_{air}\mathbf{H}\, (A - a) \qquad (47.8)$$

where a is the cross-sectional area of the iron core and A is the cross-sectional area of entire solenoid.

Magnetic Behavior of Iron When an initially unmagnetized iron rod is placed in a solenoid and subjected to a magnetic field whose field intensity $\mathbf{H}$ increases in regular increments, the flux density $\mathbf{B}$ (in the iron) does not vary linearly with the field intensity $\mathbf{H}$, but changes in accordance with a curve like that of *abcd* of Fig. 47.2. When a point (such as d) is reached where further *field intensity* produces no further *flux density* in the iron, the iron is said to have reached *magnetic saturation*.

When, after having reached d, the field intensity $\mathbf{H}$ is reduced, the magnetization curve does not retrace *dcba*, but takes a path like *de*, indicating that, although the magnetizing *field* has returned to zero, the iron itself still retains magnetic properties. The residual value *ae* of the flux density is termed the *retentivity* of the iron.

If the solenoid's magnetic field is now reversed and increased in the opposite (negative) direction, the magnetization curve takes a form similar to *efg*. The point g indicates saturation in the opposite direction. If the negative field intensity is now decreased (that is, made more positive), curve *gh* will result, and finally, if the field intensity is again increased in the original (positive) direction, the curve *hid* is attained. The closed figure *defghid* is termed a "hysteresis loop," and the area within the loop is a mea-

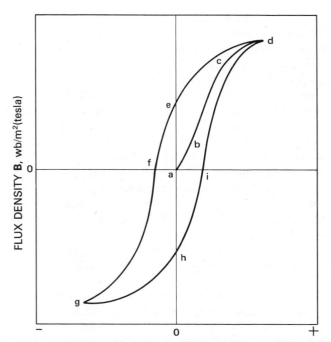

FLUX DENSITY **B**, wb/m²(tesla)

FIELD INTENSITY (MAGNETIZING FIELD) **H**, amp-turn/meter

Figure 47.2. Typical **B**–**H** curve (or *hysteresis* curve) for a cylindrical core rod of soft iron.

sure of the energy required to put the iron rod through the hysteresis cycle. (The word "hysteresis" is from the Greek, meaning "lagging behind.") In this context, changes in flux density **B**, in the iron, lag behind changes in field intensity **H**.

PROCEDURE AND MEASUREMENTS

Part 1. Calibration of the Ballistic Galvanometer

Solenoid with air core. The hysteresis apparatus is shown in Fig. 47.3, and a circuit diagram for setting up the appa-

ratus is indicated in Fig. 47.4. Figure 47.5 shows a suitable type of galvanometer. The resistance in box R_1 (Fig. 47.4) should be adjusted to provide for suitable galvanometer deflections. Begin with switches 1, 2, 3, and 4 all open and the reversing switch K_3 closed (either side). Close K_1 to the galvanometer. Then close K_2 and note the galvanometer throw and ammeter reading. Adjust R_1 and R_2 (see Fig. 47.6) until the galvanometer and ammeter readings are suitably "on scale," and then do not change these resistances again during the present test. This entire Part 1 is for the purpose of calibrating the galvanometer. Enter all Part 1 readings in Table 47-1 of the Data Sheet. After the galvanometer has returned to zero, open K_2 and record the galvanometer throw again. Then close switch 1 on the apparatus and then switch K_2, recording the galvanometer throw and ammeter reading again. After the galvanometer returns to zero, open switch K_2 and note the galvanometer throw.

Now close switches 2, 3, and 4 in turn, taking this same set of readings each time. Five sets of readings will now have been recorded. Count and record the number of turns N on the primary coil P; and measure L, the length of the solenoid. From these data calculate, from Eq. (47.1), the value of the field intensity **H**, at the center of the solenoid, for each value of the current I.

Measure the mean diameter of the primary coil P, and calculate its cross-sectional area A. Using Eq. (47.7) calculate the total flux Φ in the air-core solenoid for each value of the current I. Plot a smoothed curve on graph paper (Φ values on the vertical axis and galvanometer throws, d, on the horizontal axis) to calibrate the galvanometer in terms of flux changes within the solenoid.

Part 2. Demagnetizing the Iron Rod

The test must begin with a sample of iron which has no residual magnetism, that is, one in which all the magnetic "domains" are arranged at random. Demagnetizing is

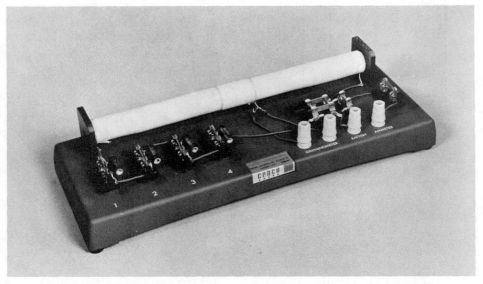

Figure 47.3. Long solenoid assembled with other components required for a hysteresis test. *(Cenco Scientific Co.)*

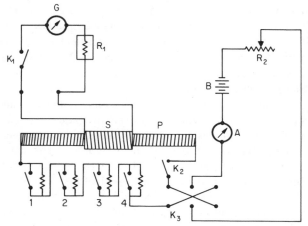

Figure 47.4. Wiring diagram for hysteresis test, showing solenoid (*P*—primary coil; *S*—secondary coil), galvanometer G, battery B, ammeter A, reversing switch K_3, and resistances 1, 2, 3, and 4, with their switches.

most simply accomplished by placing the iron rod in a long, hollow solenoid which can be operated from a *low-voltage alternating current*, like that from a 12-V transformer. Slowly withdraw the iron rod from the solenoid while the alternating current is energizing it. Lay the iron rod several feet away from other iron and from electrical circuits until its use is needed in the next part of the experiment.

Part 3. Studying Hysteresis and Plotting the Hysteresis Loop

In order to obtain a series of values of **B** and **H** and not allow them to reduce to zero after each reading, it is necessary to proceed in a step-by-step manner, measuring the increments of change in magnetization produced by suc-

cessive changes in current in the coil. Adjust R_2 (Figs. 47.4 and 47.6a) so that a heavy enough current can be drawn from the battery to result in eventual magnetic saturation of the iron rod. (The instructor will suggest the proper value for the coil-rod combination in your laboratory.) Open all switches and insert the demagnetized iron rod in the solenoid. (*Carefully* perform the following steps, in *sequence*.) Close the reversing switch K_3 (either way). Close K_1, and then K_2, and record the galvanometer throw, *as the switch is closed*, and the ammeter reading. Now close switches 1, 2, 3, and 4 *in sequence*, reading and recording the galvanometer throw and the ammeter reading as each is closed. Allow the galvanometer to return to zero from one step, before going on to the next step.

Now take five more sets of readings by *opening* the switches in reverse order 4, 3, 2, 1, and finally the switch K_2.

Throw the reversing switch to the opposite side, which reverses the current direction in the solenoid and therefore reverses the direction of the magnetic field. Record the galvanometer throw and the ammeter reading as the K_2 switch is closed to begin the second part of the study. Repeat all the above readings in sequence with the current

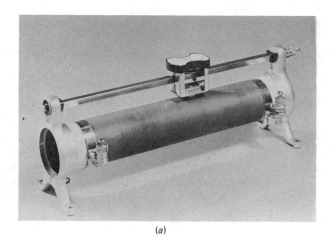

(a)

(b)

Figure 47.6. (*a*) Rheostat of a type suitable for use as R_2 in the circuit of Fig. 47.4. (*b*) Decade resistance box of type suitable for use as R_1. *(Leeds and Northrup)*

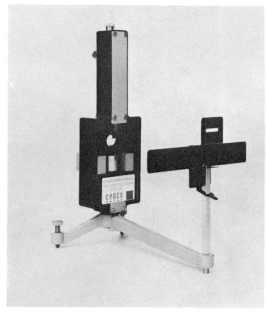

Figure 47.5. Galvanometer of a type suitable for use with the hysteresis apparatus. *(Central Scientific Co.)*

changing from zero (before K_2 is closed) to negative maximum, and back to zero. The last reading of this sequence is obtained as K_2 is opened.

Finally, throw the reversing switch again to the positive side, and take a set of five readings from zero current to the maximum again.

Reminder: The numbered switches and K_2 *must be opened and closed in proper sequence*. One misstep means starting all over again, since a departure from the sequence would change the pattern of magnetization of the "domains" and readings from that point on would not give results which would produce points on a typical hysteresis curve. You cannot, therefore, repeat a single reading. If a mistake is made, the entire cycle will have to be repeated, *beginning with the demagnetization of the iron rod* all over again.

There will be 25 sets of readings in all. Use a data sheet like the sample shown in Table 47.2, and record everything carefully. Measure the diameter of the iron rod and compute its mean cross sectional area, a.

CALCULATIONS

Using the galvanometer calibration curve prepared from Part 1 above, obtain, by interpolation, the values of changes in flux $\Delta\Phi$ for each step in the cycle of readings. Enter these (in Table 47.2) of the Data Sheet, in the column headed $\Delta\Phi$. The *total flux change* Φ is obtained by adding *algebraically* the separate $\Delta\Phi$ values. In other words, $\Phi = \Sigma\Delta\Phi$.

Compute **H** for each step (amperes $\times$ turns per meter) and the value of **H** $(A - a)$ for each step. The entries in the column headed Ba can then be obtained from the use of Eq. (47.8). Values of **B** and μ can now be calculated.

By plotting corresponding values of *flux density* **B**, and *intensity of the magnetizing field* **H**, for the 25 sets of readings, plot the complete hysteresis loop for the iron rod you used.

ANALYSIS AND INTERPRETATION

1. In an unabridged dictionary look up the word "hysteresis." Show how its meaning in the original Greek applies to this experiment.

2. Explain why alternating current demagnetizes iron.

3. In a handbook, study the hysteresis curves for steel and bronze. How do they differ from that of soft iron? Could you infer anything from the hysteresis curves of these three materials that would indicate which of the three would be the best material for use in constructing the rotors of induction motors and electric generators?

4. Hysteresis is not limited to magnetic phenomena. Read about *elastic* hysteresis, related to stress and strain in solid materials, and sketch a typical stress-strain diagram that illustrates elastic hysteresis. Can you think of an example of a practical application in which elastic hysteresis is involved?

NOTES, CALCULATIONS, OR SKETCHES

DATA SHEET

EXPERIMENT 47 ■ MAGNETIC INDUCTION IN AN IRON ROD— HYSTERESIS

Table 47.1. Suggested Format for Recording Galvanometer Calibration Information (Entries Are Illustrative Only—Not Intended to Suggest the Values You Should Obtain)

Switch	Gal. defl. (sw. closed) (cm)	Gal. defl. (sw. open) (cm)	Mean defl. of gal. (cm)	I (amp)	H amp-turn per meter	$\mu_{air}HA = \Phi$ (Wb)
K_2	0.50	0.52	0.51	0.15		
1	1.10	1.15	1.12	0.45		
2	Etc.			0.90		
3				1.30		
4				1.90		

N (turns on *primary* (*P*) of solenoid) _____

A (cross-sectional area of solenoid) _____ m^2

Table 47.2. Suggested Format for Arranging Data and Calculations for Hysteresis Study (Entries are Illustrative Only.)

Switch	Galv. throw (cm)	Flux change, $\Delta\Phi$ (Wb)	Tot. flux, $\Phi = \Sigma\Delta\Phi$ (Wb)	Current, I (A)	Field intens. (H)	H(A − a)	Ba	Flux dens. (B)	$\mu = B/H$
K_2, closed	2.4			0.15					
1, closed	7.6			0.45					
2, closed	etc.			0.90					
3, closed				etc.					
4, closed									
3, opened									
2, opened									
1, opened									
K_2, opened									
Reverse, etc.									

a (cross-sectional area of the iron core) _____ m^2

PRINCIPLES OF GENERATORS AND MOTORS

PURPOSE

To study the principles of operation of simple dc and ac generators and motors.

APPARATUS

St. Louis motor with both dc and ac rotors and separate field winding; bar magnets; galvanometer; source of 6-V direct and 6-V alternating current; rheostat; small compass; wire for connections.

INTRODUCTION

Almost 80 percent of today's work (in the United States)—in the factory, on the farm, or in the home—is accomplished by electrically powered machinery. Electric motors do the work, and electric generators provide the electric energy required by the motors. A *generator* is a device for converting mechanical energy into electric energy, and a *motor* is a device to perform the opposite function.

Actually, motors and generators are basically much the same as far as their construction is concerned. In fact, they are often *reversible*. That is, a generator will run as a motor if electric energy is supplied to it, and if mechanical energy is supplied, it will produce electric energy.

The parts of simple motors and generators are listed and defined as follows (refer to Figs. 48.1 and 48.2):

1. The *rotor*—the rotating part, wound with conductors.
2. The *stator*—the stationary part, also wound with conductors.
3. The *armature*—that part in which the induced currents are produced and from which they are collected (or the rotor of a dc motor).
4. The *field winding*—that part which produces the magnetic field; it may be on either stator or rotor, depending on the type of machine.
5. *Slip rings*—devices for collecting the alternating current produced in generator rotors.
6. The *commutator*—a device for collecting direct current from a dc generator rotor or for supplying direct current to the rotor of a dc motor.
7. *Brushes*—sliding contact devices to pick up the generated current from armature-rotor machines, or to supply current to dc motors.

Part I. Dc Machines

Dc generators are ordinarily of the *armature-rotor field-stator* type. The field windings are on stationary iron cores, and a steady dc voltage is applied to them, creating a magnetic field in which the rotor turns. The rotor itself is wound with conductors in which the induced emf is produced. The rotor is therefore the armature in this case. The induced currents are led to the external circuit by means of the *commutator* and *brushes*.

Dc motors are also of the armature-rotor field-stator type. Current from an external source is supplied directly

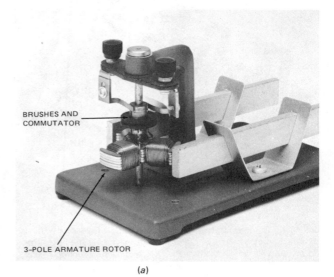

BRUSHES AND COMMUTATOR

3-POLE ARMATURE ROTOR

(a)

SLIP RINGS

(b)

Figure 48.1. St. Louis machines operating as generators. (*a*) A three-pole dc generator. Note the collecting brushes on the armature rotor. Bar magnets make up the field stator. (*Central Scientific Co.*) (*b*) A two-pole ac generator. The armature rotor is fitted with slip rings. The field stator is made up of bar magnets.

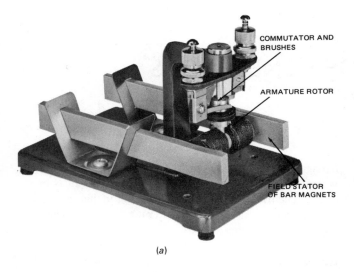

Figure 48.2. St. Louis motors for laboratory study. (*a*) Motor with dc armature rotor and commutator, and with bar magnets as the field stator. (*b*) Same, but with an electromagnetic-type field stator. (*Central Scientific Co.*)

to the field windings (stator) and via the brushes and commutator to the armature (rotor). Two separate magnetic fields are thus produced, and the interaction between these fields results in the turning of the rotor in its bearings.

Part 2. Ac Machines

Small-sized *ac generators* are mostly of the armature-rotor field-stator type. Large machines, however, such as those used in the commercial production of electric power, are almost invariably of the armature-stator field-rotor type. Direct current is fed to the rotor in these large machines, and a *rotating magnetic field* is thus produced. The armature windings are on the stator, and the electric energy generated may thus be collected and led away from the machine without resort to sliding contact devices such as brushes.

Ac motors, except for small-sized household appliance types (1/16 hp and less), are often of the *induction* type. A polyphase ac emf is supplied to the field windings (stator). A rotating magnetic field results within the region where the rotor is cradled in its bearings. No current at all is

supplied to the rotor, but the rotating magnetic field, as it cuts across the conductors with which the rotor is wound, *induces* a current in the rotor windings. The interaction between the two fields thus produced causes the rotor to turn. Such motors are often referred to as "squirrel cage" motors.

Ac induction motors will not be studied in this experiment since the equipment required is ordinarily not available in the physics laboratory.

In the observations to follow, take careful notes on every step. Draw sketches to illustrate the flow of current and the interaction of magnetic fields. These notes and sketches constitute your Data Sheet, so make them neatly and carefully as you proceed. We will make use of greatly simplified "models" of generators and motors. They actually operate, however, and are so constructed that study of generator and motor principles is easily accomplished (Figs. 48.1 and 48.2).

MEASUREMENTS AND OBSERVATIONS

Part I. Dc Machines

A. Simple Dc Generator (Refer to Fig. 48.1*a*).   Use the dc rotor, i.e., the one with the commutator. See that the rotor turns freely. If it does not, adjust the bearings and brushes. Add a drop of light oil if needed. Set the bar magnets close to the rotor. Connect the galvanometer to the binding posts on top of the brush holder. Rotate the rotor (armature) by hand in one direction, and note direction of galvanometer deflection. Rotate in opposite direction, and note galvanometer deflection. Note the effect of different speeds of rotation. Move bar magnets further away from rotor, and note the effect on the induced emf.

Replace the bar magnets by the electromagnetic field winding. "Excite" the field (2 or 3 V from the laboratory dc source), and repeat the above.

Sketch circuit diagrams, showing the direction of the magnetic field, of the rotor coil winding, and of the rotation. By the generator (three-finger) rule, show the direction of induced current.

Show by diagrams and sine-wave sketches that you understand the function of the commutator as used on a dc generator. Exactly how does the output current become *direct* current?

B. Simple Dc Motor

1. Bar magnet "field" (Fig. 48.2*a*). Connect a 2- or 3-V dc source in series with the armature rotor and switch. Set the brushes so that polarity of the armature changes at the right time to produce continuous rotation. Test with a small compass the polarity of the armature core for one complete revolution.

Investigate the following.
a. Test the effect on the speed of the motor of moving the magnets farther from the armature.
b. Test the effect on speed of reducing armature current, by use of a rheostat.
c. How would you reverse the direction of rotation? (Re-

verse armature current? Reverse magnetic field? Reverse both?) Test all these possibilities.

2. Electromagnet "field" (Fig. 48.2*b*). Take the bar magnets out, and replace them with the electromagnet. Excite the field from one source, and supply the rotor from a different source (both about 2 V). Reverse the current in the field, then in the armature, and finally in both. Explain the effects observed.

Connect a sensitive ammeter in the rotor circuit. Note its reading when the rotor is held stationary. Then allow the motor to run, and note the ammeter reading. Explain your observations.

Series-Wound Motor   Connect the armature and the field in series from the *same* source. What effect does reversing the current direction have on the direction of rotation?

Shunt-Wound Motor   Connect the armature and the field in parallel from the same source. Again, what effect does reversing the current direction have? Experiment with loading these two types of motors. That is, make them do some external work. In your report discuss at some length the relative advantages of series- versus shunt-wound motors.

Part 2.   Ac Machines

A.   Simple Ac Generator   Replace the dc rotor with the ac rotor and slip rings (Fig. 48.2*b*). Excite the *field stator* from the laboratory dc source (about 1 V). Connect the galvanometer to the rotor, and study the nature of the ac output while turning the rotor by hand. This arrangement represents an *armature-rotor-field-stator* machine.

Then excite the *rotor* from the laboratory dc source, and connect the galvanometer to the stator. Note the character of the emf produced. This arrangement represents a *field-rotor-armature-stator* machine.

B.   Simple Ac Motor   Put the dc rotor (with commutator) back in the machine. Connect the field magnet and rotor in *series* with a 6-V ac source from a transformer. Run the motor. Describe its operation. What type of motor will run on either direct or alternating current?

Part 3.   Motor-Generator Set

Set up the dc machine and the ac machine as in Fig. 48.3. Run the motor from a laboratory source (2–3 V), and observe the output from the ac generator.

DATA SHEET

Prepare your own Data Sheet. It should include complete notes and sketches on the observations.

PRESENTATION OF RESULTS

Divide this section into the separate headings of the Measurements and Observations section. Answer all the questions expressed and implied in that section. Make careful drawings to illustrate the findings.

ANALYSIS AND INTERPRETATION

In separate paragraphs discuss what you have learned about simple dc and ac generators and motors. In addition, answer the following:

1. By means of a diagram explain why an emf curve (plotted against electrical degrees) is a sine wave.
2. Give the principal reasons why most large generators are:
a. Alternators, rather than dc generators
b. Of the field rotor, armature-stator type
3. Indicate the function of:
a. A commutator on a small dc generator
b. A commutator on a simple dc motor
c. Slip rings on a small ac generator

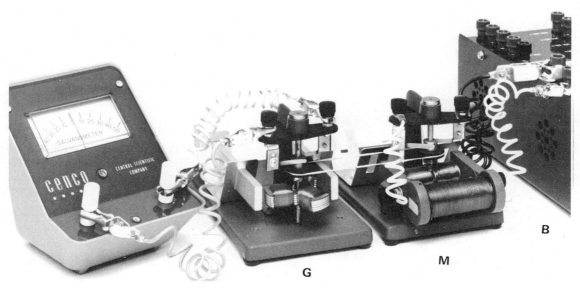

Figure 48.3.   Model of a simple motor-generator set, using the simplified St. Louis machines. Battery *B* energizes dc motor *M*, which, by means of a string-belt drive, turns three-pole dc generator *G*. Output emf of generator is observed on the galvanometer at left. *(Central Scientific Co.)*

d. Brushes on a large alternator

4. Explain how an induction motor can operate without having any electric current supplied to its rotor.

5. Explain what is meant by ''back emf'' in motor windings. What effects does it have?

6. Why is it common practice to construct the iron cores of rotors and stators of thin sheets—laminations—rather than of solid iron?

7. Name four types of ac motors and describe how each operates.

NOTES, CALCULATIONS, OR SKETCHES

EXPERIMENT 49

PROPERTIES OF ALTERNATING-CURRENT SERIES CIRCUITS

PURPOSE

To study simple ac series circuits involving resistance, inductance, and capacitance; and to investigate the effects of variations in these factors on the power in ac circuits.

APPARATUS

Cenco impedance apparatus or equivalent; digital multimeter or vacuum-tube voltmeter; ac ammeter (0- to 0.5-amp range); 350-Ω rheostat; connecting wires and probes. Fig. 49.1 shows two forms of the apparatus. Either form

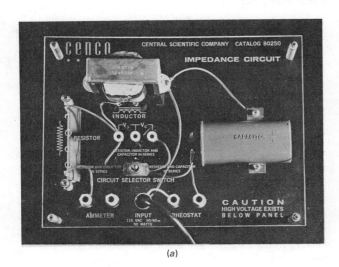

(a)

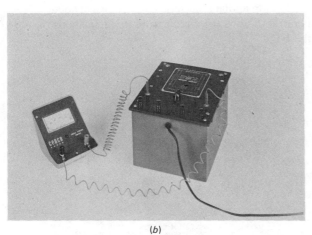

(b)

Figure 49.1. Two forms of impedance apparatus for studying the properties of ac circuits in the laboratory. (a) Student form, showing all the elements of the circuit; (b) enclosed form, with circuit diagram imprinted on the top cover. Both require an ac ammeter, a variable rheostat, and a sensitive ac voltmeter such as a vacuum-tube voltmeter (VTVM) (or a digital multimeter), as associated equipment. (Central Scientific Co.)

may be used, or apparatus for the purpose may be assembled locally.

INTRODUCTION

The factors which affect energy flow in ac circuits are *current, emf,* and *resistance to electron flow*. The "resistance" of an ac circuit, however, is a more complex concept than the ohmic resistance of dc circuits. Three properties of ac circuits affect the overall resistance (called *impedance*)—*ohmic resistance, capacitance,* and *inductance,* abbreviated R, C, L.

In an ac circuit containing only *ohmic resistance,* current and voltage are in phase and the power in the circuit (watts) equals the product of volts times amperes:

$$P = V_{rms} I_{rms} \qquad (49.1)$$

where the designation *rms* refers to the *root mean square* (or *effective*) values of voltage and current. (Ac voltmeters and ammeters read directly in *rms* values.) *In all the discussion and equations below, rms (effective) values of current and voltage are implied.*

If only *capacitance* is present, the buildup of *voltage* is delayed and voltage lags current by 90 electrical degrees. The resistance to current caused by capacitance is called *capacitive reactance.* It is measured in ohms and is given by the formula

$$X_C = \frac{1}{2\pi f C} \qquad (49.2)$$

when f is in cycles/s (Hz) and C is in *farads* (F).

If only *inductance* is present, the buildup of *current* is delayed and current lags voltage by 90 electrical degrees. The resistance to current caused by inductance is called *inductive reactance,* which is also measured in ohms. It is given by the formula

$$X_L = 2\pi f L \qquad (49.3)$$

when f is in cycles/s (Hz) and L is in henrys (H).

Ac circuits rarely, if ever, contain just one of the *RCL* factors. Two, or frequently all three of them, are involved. The net total resistance to electron flow is the *impedance* Z, and Ohm's law for ac circuits becomes

$$I = \frac{V}{Z} \qquad (49.4)$$

Vector methods simplify the explanation of *RCL* relationships when resistance, capacitance, and inductance are in series in an ac circuit. R alone does not result in a phase

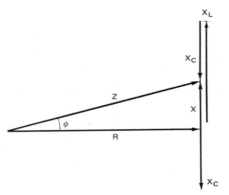

Figure 49.2. The vector relationships that exist among ohmic resistance, inductive reactance, capacitive reactance, total (net) reactance, impedance, and phase angle in ac circuits.

displacement. C alone gives a 90° phase shift (current leads voltage), and L alone gives a 90° shift in the opposite manner (voltage leads current). Vectors are used to portray RCL relationships as follows (see Fig. 49.2).

The ohmic resistance vector **R** is first plotted horizontally to scale. The inductive reactance vector $\mathbf{X}_L$ is then plotted from the end of the **R** vector upward at 90°. The capacitive reactance vector $\mathbf{X}_C$ is plotted in the opposite direction, vertically downward (180° from $\mathbf{X}_L$). The vector sum of $\mathbf{X}_L$ and $\mathbf{X}_C$ is **X**, the net *reactance*. The vector sum of **R** and **X** is **Z**, the *impedance*.

From the vector diagram (Fig. 49.2), there results the vector equation:

$$\mathbf{X} = \mathbf{X}_L + \mathbf{X}_C$$

Or, in scalar terms,

$$X = X_L - X_C$$

and, vectorially

$$\mathbf{Z} = \mathbf{R} + \mathbf{X}$$

Algebraically,

$$\begin{aligned} Z &= \sqrt{R^2 + X^2} \\ &= \sqrt{R^2 + (X_L - X_C)^2} \\ &= \sqrt{R^2 + \left(2\pi fL - \frac{1}{2\pi fC}\right)^2} \end{aligned} \qquad (49.5)$$

The angle ϕ of Fig. 49.2 is known as the *phase angle*, and it tells to what extent voltage and current are out of phase. If ϕ is above the R axis, X_L is greater than X_C and voltage leads current by ϕ electrical degrees. If ϕ is below the R axis, X_C is greater than X_L and current leads voltage by ϕ electrical degrees.

If $X_L = X_C$ (the condition of *series resonance*), $\phi = 0$ and voltage and current are in phase. When this condition obtains, the power in an ac circuit is maximum, and is simply the product of rms current and rms voltage, given by Eq. (49.1).

When X_L and X_C are unequal, the *true power* in the circuit is less than the *apparent power* obtained from the product VI. Since $V \cos \phi$ is the component of V that is in

phase with I, the effective power actually used in an ac circuit is given by

$$\text{True power, } P = IV \cos \phi \qquad (49.6)$$

From Fig. 49.2, $R/Z = \cos \phi$. The cosine of the phase angle is called the *power factor*.

$$\text{Power factor (p.f.)} = \cos \phi = \frac{R}{Z} \qquad (49.7)$$

or, power factor in an ac circuit is equal to the ratio of its ohmic resistance to its impedance.

In words,

$$\begin{aligned} \text{Power (watts)} = \\ \text{volts (effect.)} \times \text{amperes (effect.)} \times \text{p.f.} \quad (49.8) \end{aligned}$$

Note that actual (true) power *used* is expressed in watts (W) or kilowatts (kW). *Apparent power,* supplied to the line by a power company, is ordinarily expressed in volt-amperes (VA) or kilovolt-amperes (kVA).

MEASUREMENTS

Connect the rheostat and the ammeter to the impedance apparatus across the terminals indicated on top of the box (see Fig. 49.1). (The instructions following are keyed to the apparatus of Fig. 49.1*b*.) If a different apparatus is used, follow the directions of its supplier.)

Part I. *RCL* Circuit

Set the selector switch on the apparatus so that R, C, and L are all in a series circuit. Plug the impedance apparatus into the ac line current. Adjust the rheostat to give a current of about 0.3 amp. Then, with the vacuum-tube voltmeter (VTVM) or digital multimeter, (see Fig. 49.3), measure the voltage drop across R, C, L, and Z in turn. Record in the Data Table. For a second trial, increase the current to 0.325 amp and again measure the potential differences across R, C, L, and Z. Repeat in steps of 0.025 amp until a final series of readings at about 0.6 amp is made. Sketch the circuit from the indications on top of the box.

Figure 49.3. One type of vacuum-tube voltmeter. *(Central Scientific Co.)*

Part 2. RL Circuit

Short out the *capacitor* with the selector switch, and take a series of measurements similar to those in Part 1, but involving only R, L, and Z.

Part 3. RC Circuit

Short out the *inductance* with the selector switch, and take a third series of measurements involving only R, C, and Z.

PRESENTATION OF RESULTS AND CALCULATIONS

1. RCL circuit

a. On a sheet of graph paper lay out horizontally a scale of current readings. Voltage drop readings are to be plotted along the vertical axis. On the same sheet, plot four curves, one for the voltage drop across the resistor, one for that across the capacitor, one for that across the inductor, and one for that across the entire circuit. Label them.

b. Note that the C, Z, and R curves are essentially straight lines, whereas the L plot is a curve. (This is because the degree of magnetization of an iron core varies with the current applied, or with the extent of magnetic "saturation" of the core (see Experiment 47).

c. From Eq. (49.4), $Z = V/I$, the impedance of each component and of the total circuit in series may be obtained from the slope of the plotted "curves." (The "slope" of a curve at any point is the slope of the tangent at that point. Examination of your plot will reveal that the value of the tangent in each case is the ratio V/I.). For the inductance curve, draw a tangent at the 0.4-amp line, and use the slope of this tangent to determine Z. From such an analysis, get answers for R, X_L, X_C and Z.

On another sheet of graph paper plot a *vector* diagram using the calculated values of R, X_L, and X_C. Solve vectorially for **Z,** and compare this answer with that obtained in (c) above.

2. Make the same kind of a plot for the *RL* circuit, using data from Part 2. Calculate R, X_L, and Z. Plot a vector diagram, using the calculated values of R and X_L. Solve vectorially for **Z** and compare with the calculated value.

3. Repeat for the *RC* circuit.

4. Remembering that $X_L = 2\pi fL$ and that $X_C = 1/2\pi fC$, for the value of $I = 0.4$ amp, calculate the value of L in millihenrys (mH) and the value of C in microfarads (μF) for a frequency of 60 Hz. Present all results in a neat summary table.

ANALYSIS AND INTERPRETATION

In this section be sure to include a discussion of the following:

1. Why is the voltage drop across the capacitor alone greater than that across the entire circuit?

2. What is meant by *series resonance* in ac circuits?

3. Define phase angle and show how it can be found from a vector diagram.

4. Express "power factor" in terms of a vector diagram and in terms of true and apparent power.

5. How can the power factor of an inductive circuit be improved? Explain using diagrams.

6. Why is "power on the line" usually expressed in kilo-volt-amperes (kVA) by power companies rather than in kilowatts (kW)?

7. Give a brief discussion of the phenomenon of electrical resonance. Relate it to the other kinds of "resonance" encountered in previous studies in physics. What unifying concept is always present whenever resonance in any form occurs?

NOTES, CALCULATIONS, OR SKETCHES

Name _____ Date _____ Section _____

Course _____ Instructor _____

DATA SHEET

EXPERIMENT 49 ■ PROPERTIES OF ALTERNATING-CURRENT SERIES CIRCUITS

Part 1. *RCL* Circuit

Current (I) A	Voltage drop (V) across components			
	R	C	L	Z

Part 2. *RL* Circuit

Current (I) A	Voltage drop (V) across components,		
	R	L	Z

Part 3. *RC* Circuit

Current (I) A	Voltage drop (V) across components		
	R	C	Z

PART SEVEN ■ MODERN PHYSICS

EXPERIMENT 50

THE SEMICONDUCTOR DIODE CURRENT-VOLTAGE CHARACTERISTICS

PURPOSE

To obtain experimental data for the relationship between current and voltage for germanium and silicon semiconductor diodes, and to produce a graphical display of the current-voltage relationships.

APPARATUS

Germanium diode (1N299 or similar); silicon diode (1N645 or similar); 1 KΩ and 10 KΩ resistors; 10 KΩ potentiometer; adjustable dc power supply (0–20 V); 9.0-V battery; digital multimeters; solderless breadboard for constructing circuits; wires for connections.

INTRODUCTION

A semiconductor diode is a two-terminal device that has a resistance that depends on both the magnitude and the polarity of the voltage applied to its terminals. A diode, usually made of the semiconducting element silicon, contains *p*-type (positive) and *n*-type (negative) regions as illustrated in Fig. 50.1*a*. The impurity atoms that are added to the silicon to make it a *p*-type semiconductor have one fewer valence electrons than silicon atoms have. As a result, the *p*-type material contains many *holes* that behave like free positive charges. Similarly, the impurity atoms that are added to the silicon to make it an *n*-type semiconductor have one more valence electron than silicon atoms have. As a result, the *n*-type material contains many *free electrons*.

At the junction between the *p*- and *n*-sides of the diode there is a narrow region that contains essentially no free charges. In that region, called the *depletion region* or the *space-charge region,* there is an electric field that points from the *n*-side to the *p*-side of the junction. When no voltage is applied to the diode (*zero bias* condition), the width of the depletion region and the strength of the electric field in the depletion region automatically adjust so that the net current across the junction is zero. Under the zero bias condition, the current of electrons that *diffuse* from the *n*-side to the *p*-side of the junction just equals the *drift* current of thermally generated free electrons from the *p*-side to the *n*-side of the junction. Similarly, the diffusion current of holes from the *p*-side to the *n*-side of the junction is equal to the drift current of thermally generated holes from the *n*-side to the *p*-side of the junction.

A *p-n* junction diode is said to be *forward biased* when a voltage is connected to the diode as shown in Fig. 50.2*a*. When the applied voltage has the opposite polarity, the diode is said to be *reverse biased* (see Fig. 50.2*b*). When a diode is forward biased, the width of the depletion region and the strength of the electric field in the depletion region are both reduced, and therefore it becomes easier for electrons and holes to diffuse across the junction. This results in a *forward current* in the direction shown in Fig. 50.2*a*.

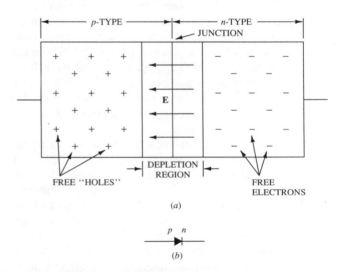

(a)

(b)

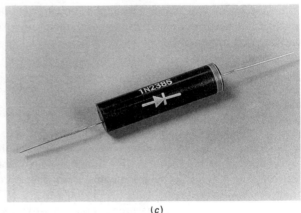

(c)

Figure 50.1. (*a*) Pictorial diagram of a *p-n* junction diode. On the *p*-side of the junction there are far more free holes than free electrons, and on the *n*-side of the junction there are far more free electrons than free holes. The electric field in the depletion region is due to the presence of negative impurity ions in the *p*-region and of positive impurity ions in the *n*-region. (*b*) Schematic symbol for a *p-n* junction diode. Although the *p*- and *n*-sides of the device are indicated in this figure, they are not usually labeled in schematic drawings that use this symbol. (*c*) A typical *p-n* junction diode.

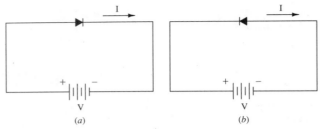

Figure 50.2. Polarity of applied voltage (*a*) to forward bias, and (*b*) to reverse bias a diode.

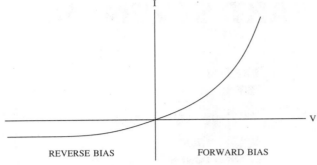

Figure 50.3. *I-V* characteristic for a typical *p-n* junction diode.

When a diode is reverse biased, the width of the depletion region and the strength of the electric field in the depletion region are both increased, and the diffusion current of electrons and holes across the junction is essentially reduced to zero. Except for very small values of reverse bias, the *reverse current* is due only to a drift current of thermally generated electrons from the *p*- to the *n*-side and of thermally generated holes from the *n*- to the *p*-side of the junction.

The movement of holes and electrons in both directions across the junction between the *p*- and *n*-sides of a diode results in a net conventional current across the junction. The manner in which the junction current depends on the magnitude and direction of the voltage applied to the diode can be displayed graphically as shown in Fig. 50.3. Such a graphical display of the relationship between current (*I*) and voltage (*V*) for a diode is usually referred to as the *I-V* characteristic for the diode.

One goal of this experiment is to determine the current-voltage relationship for a typical semiconductor diode. A secondary goal is to gain some general insight into the properties of a *p-n* junction since essentially all semiconductor devices contain one or more *p-n* junctions of the type discussed here for a diode. Some specific examples of such devices are: zener diodes, varactor diodes, tunnel diodes, light-emitting diodes, solar cells, bipolar junction transistors, unijunction transistors, and field-effect transistors. Moreover, some integrated circuits contain over one-hundred million diodes and transistors and, therefore, over one-hundred million *p-n* junctions.

MEASUREMENTS

Part I. Forward Bias

Connect an adjustable dc power supply, a silicon diode, a 1000-Ω resistor, and an ammeter in series as shown in the circuit diagram in Fig. 50.4. A digital multimeter capable of measuring currents in the microampere range is appropriate for this experiment.

Connect a voltmeter to measure the potential difference across the diode for each of the different measured currents. See Fig. 50.5 for an example of the experimental arrangement. Measure and record the applied forward voltage and the resulting diode current for forward voltages from zero up to about 0.75 V. You may not get currents larger than 100 μA until the forward voltage is about 0.5 V.

Turn off the power supply, replace the silicon diode with a germanium diode, and measure and record the forward voltage and diode current for forward voltages from zero up to about 0.5 V. With the germanium diode you should be able to get easily measurable currents when the forward voltage is 0.25 V or smaller.

Part 2. Reverse Bias

The procedure used for Part 1 cannot be used to obtain data for the voltage across a reverse-biased diode because the resistance of the voltmeter is likely to be smaller than the very large resistance of the reverse-biased diode. Data for the magnitude of the reverse bias must therefore be obtained indirectly. This can be accomplished with the circuit shown in Fig. 50.6. Since the algebraic sum of the voltage drops around a closed loop in a circuit must be zero, the voltage V_1, the voltage V_2, and the reverse bias voltage V_D for the diode must be related by the equation $V_1 - V_2 - V_D = 0$; and therefore the potential difference across the diode can be calculated using the equation

$$V_D = V_1 - V_2 \tag{50.1}$$

The value for the diode current can be obtained from the Ohm's law relationship between the values for the voltage V_2 and the resistance R_2.

Construct the circuit shown in Fig. 50.6 using a germanium diode and obtain and record data for the values of V_1 and V_2 for values of V_1 in the range from zero to 5 V.

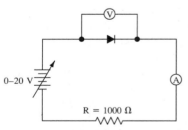

Figure 50.4. Circuit diagram for obtaining current-voltage data for a forward-biased diode.

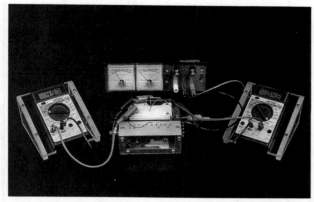

Figure 50.5. Arrangement of apparatus for the circuit of Fig. 50.4. *(MSOE Academic Media Services)*

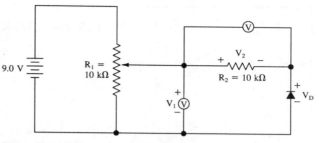

Figure 50.6. Circuit diagram for obtaining current-voltage data for a reverse-biased germanium diode.

CALCULATIONS

For each pair of values of V_1 and V_2, calculate the voltage V_D across the diode using Eq. (50.1), and calculate the diode current using the values for V_2 and R_2.

Prepare your own Data Sheet for this experiment.

PRESENTATION OF RESULTS

Prepare a plot of forward current versus forward voltage using the data for the silicon diode from Part 1. Superimpose on this graph the forward- and reverse-bias current-voltage data for the germanium diode from Parts 1 and 2. It may be appropriate to choose a more sensitive scale for plotting the reverse-bias data.

ANALYSIS AND INTERPRETATION

1. For a given value of forward bias, is the forward current greater for a silicon diode or for a germanium diode?
2. For any diode, if a sufficiently large reverse bias is applied a large reverse current will result. The value of the reverse voltage that causes the large reverse current is called the *reverse breakdown voltage*. Use your textbook or some other reference to read about the reverse breakdown voltage for a diode, and draw a sketch of an *I-V* characteristic that displays the reverse breakdown voltage and current for a diode.
3. Use your textbook or some other reference to see how semiconductor diodes are used as *rectifiers* to convert alternating current to direct current in simple power supplies. Draw schematic diagrams to show how diodes are used in *half-wave rectifier, full-wave rectifier,* and *bridge rectifier* power supplies.
4. Since the junction current for a reverse-biased diode is due essentially only to *thermally generated* free charges, it seems likely that the reverse current for a given value of reverse bias would depend on the temperature of the diode. Would you expect the junction current for a reverse-biased diode to increase or decrease when the temperature of the diode is increased?
5. Suggest a method of using a diode to monitor or detect small changes in temperature.
6. As mentioned in Question 3, diodes are used as rectifiers in power supplies to convert alternating current to direct current. Describe some other applications for diodes.

EXPERIMENT 51

THE CATHODE-RAY OSCILLOSCOPE

PURPOSE

To study the cathode-ray oscilloscope, and to become familiar with its operation and with some of its uses.

APPARATUS

Oscilloscope; audio signal generator; low-voltage, adjustable ac power supply; 1.5 V battery; digital multimeter or vacuum-tube voltmeter (VTVM); sensitive microphone; matched tuning forks with resonator boxes; various musical instruments if available; wire for connections.

INTRODUCTION

The cathode-ray oscilloscope is probably the most versatile of all electronic test instruments. It is used in almost all fields of research from electronics to nuclear engineering and biomedical technology. As a diagnostic instrument it is found serving the research scientist on the one hand and the auto mechanic on the other.

For almost all applications, the oscilloscope provides a graphical display of voltage along a vertical axis against time on a horizontal axis. The display appears on the screen of a cathode-ray tube. (See Fig. 51.1.)

The cathode-ray tube (CRT) is the heart of the oscilloscope (see Fig. 51.2). It is a rather elongated vacuum tube with a source of high speed electrons (an *electron gun*) at one end and a fluorescent screen at the other. The electron gun furnishes a supply of high-speed electrons, all having nearly the same velocity, and by means of an "electronic lens system," brings them into a narrow beam and accelerates them down the axis of the tube. If the beam is not deflected, it will strike the fluorescent screen, and a small bright spot will be seen.

Forces for deflecting the beam are supplied by electric fields established between each of two pairs of deflection

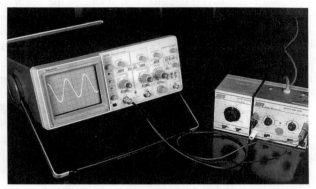

Figure 51.1. An oscilloscope displaying the sine wave output signal from an audio signal generator. *(MSOE Academic Media Services)*

plates placed in the tube so that the beam must pass through both sets on the way to the screen. One set of plates is horizontally oriented. The electric field established between these plates will be vertically directed (upward or downward), and an electron beam passing between those plates will be deflected downward or upward. The horizontally oriented plates are therefore called the *vertical deflection plates*. The other pair of plates is vertically oriented. An electron beam passing through the horizontally directed electric field between those plates will be horizontally deflected to the right or left depending on the direction of the electric field. The vertically oriented plates are therefore called the *horizontal deflection plates*. As the electron beam moves up and down or back and forth while passing through the two sets of deflection plates, its end point describes a corresponding "trace" on the screen.

In the oscilloscope, the cathode-ray tube is combined with various electronic circuits which produce the voltages

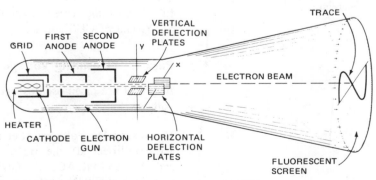

Figure 51.2. Diagram of the principal components of one type of cathode-ray tube (CRT).

and signals required for its operation. A *power supply* furnishes the necessary voltages to operate the electron gun and to accelerate and focus the beam. An amplifier which "drives" the vertical deflection plates, allows a weak input signal to produce a pronounced vertical deflection of the electron beam. Another amplifier drives the horizontal deflection plates. Each amplifier has a gain control which can be adjusted to cause the presentation on the screen to become larger or smaller.

There is a "sweep circuit" which causes the beam to sweep horizontally from left to right across the screen at a constant speed and then jump quickly back from right to left to repeat the sweep. The signal from the sweep circuit is referred to as a "sawtooth potential" because of its configuration when voltage is plotted against time (see Fig. 51.3). The repetition frequency of the sweep can be varied in order to provide a display of a desired number of cycles of the signal being applied to the vertical deflection plates. The synchronization of these two signals is accomplished by a *trigger system* which causes the sweep to start at the right time so that the resulting pattern on the screen will appear to stand still.

Cathode-ray oscilloscopes (CROs) are available that can analyze, diagnose, and record rapidly changing electrical phenomena whose frequencies may vary from less than 1 Hz to thousands of MHz. Potentials of less than 1 microvolt (μV) can be studied, and time intervals of less than 1 nanosecond (ns) can be measured.

In the investigations to follow, the first steps involve familiarization with the main oscilloscope controls. Following that exercise are several elementary studies which serve to show the versatility of the instrument. *Warning!* High voltages are present in a CRO. Observe all safety precautions.

MEASUREMENTS

Part I. Familiarization with the Instrument

Although the details of the front panel arrangement of controls may vary with the make of the instrument, every oscilloscope has three main groups of controls:

1. Controls to adjust the appearance and position of the trace. The INTENSITY control allows the brightness of the trace to be varied, and the FOCUS control can be used to adjust the sharpness of the trace. The horizontal and vertical POSITION controls can be used to change the location of the trace on the screen.

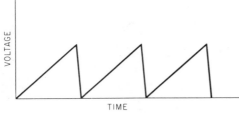

Figure 51.3. The sawtooth potential output from the sweep generator circuit of an oscilloscope.

2. Controls associated with vertical deflection. Coarse and fine gain controls allow selection of the level of amplification of the signal applied to the vertical input. The coarse gain control is usually labeled VOLTS/DIV or VOLTS/CM. The fine gain control must be in the CAL (calibrated) position in order for the numbers associated with the positions of the coarse gain control to be valid. A switch labeled AC, DC, GND (ground) is usually located near the gain control. Oscilloscopes with two vertical input channels will have two sets of these controls, one set each for channel 1 and channel 2. Channels 1 and 2 are sometimes labeled X and Y.

3. Controls associated with horizontal deflection. Coarse and fine controls allow adjustment of the frequency of the sawtooth oscillator in the sweep generator circuit. The coarse control is usually labeled TIME/DIV or SEC/DIV to indicate the time for the electron beam to move horizontally one division across the face of the cathode-ray tube. One position of the coarse frequency control should allow an EXTERNAL signal, rather than the sawtooth signal, to be applied to the horizontal deflection plates. The fine frequency control must be in the CAL (calibrated) position in order for the numbers associated with the positions of the coarse frequency control to be valid. Single channel oscilloscopes will also have a HORIZONTAL gain control.

Controls for the trigger system circuits usually allow selection of INT (internal) (or CH1 or CH2 for dual trace oscilloscopes), LINE (60-Hz signal), or EXT (external) sources of triggering signals. In addition, switches labeled AUTO or NORM (normal) allow selection of different modes of triggering.

Plug the power cord into the 115-V ac line, turn on the instrument, and set the TIME/DIV switch in the EXT (or XY) position. When the beam appears after about a minute, adjust the INTENSITY and FOCUS controls until a sharply defined but relatively faint spot is seen on the screen. (NOTE: Do not allow a spot of high intensity and small area to remain stationary on the screen for any appreciable length of time, as it may "burn" or discolor the screen material.)

Move the spot to the center of the screen using the POSITION controls. Set the trigger mode switches to AUTO and INT, and the TIME/DIV switch to one of the lower frequencies (larger values of TIME/DIV) provided by the internal sweep generator. Turn up the HORIZONTAL gain control until the beam traces a path across the full width of the screen. Make final POSITION, INTENSITY, and FOCUS adjustments.

Part 2. Measuring Voltages

A typical use of an oscilloscope is to measure dc and ac voltages.

Measuring Dc Voltages Set the TIME/DIV switch to 1 ms (millisecond), the trigger mode switches to AUTO and INT, and the AC, DC, GND switch to DC. Use the vertical POSITION control to position the trace on the

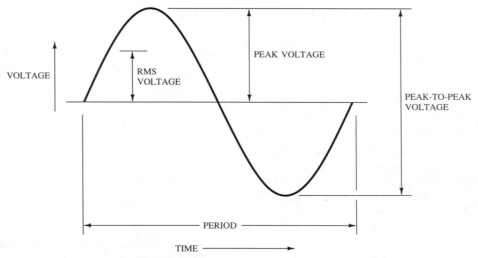

Figure 51.4. Diagram showing the relationship between *peak, peak-to-peak,* and *rms* (root-mean-square) voltages for a voltage that varies sinusoidally with time. The *period* of the waveform is the time required for one complete cycle.

center line of the graticule of grid lines on the screen. With the VOLTS/DIV switch set to 0.5 V, connect a 1.5-V battery to the vertical input with the negative terminal of the battery connected to the ground side of the input connector. Check to see if the trace moves upward approximately 1.5 divisions. Change the polarity of the battery connection to the vertical input and note that the trace deflection occurs in the downward direction.

Measuring Ac Voltages   Set the TIME/DIV switch to 2 ms, and the AC, DC, GND switch to AC. Use the vertical POSITION control to position the trace on the center line on the screen. Set the control of an adjustable ac power supply to provide an output voltage of several volts. Connect the output of the power supply to the vertical input of the oscilloscope and select a position for the VOLTS/DIV switch to provide a display of the sine wave of a size that nearly fills the screen. Determine the peak-to-peak voltage of the sine wave (see Fig. 51.4).

Disconnect the output of the ac power supply from the oscilloscope and connect it to a digital multimeter (or VTVM) set to measure ac voltages. Since the multimeter is designed to display an rms (root-mean-square) value for the ac voltage (see Fig. 51.4), the value displayed for the voltage will be less than the peak-to-peak value obtained using the oscilloscope. The rms voltage should be approximately equal to the peak-to-peak voltage divided by $2\sqrt{2}$. Check your values to see if this relationship holds for your measurements.

Part 3.   Measuring Frequency

Set the frequency control of a sine-wave audio signal generator (Fig. 51.5) to 1000 Hz and connect the output of the signal generator to the vertical input of the oscilloscope. Adjust the TIME/DIV switch so that at least one cycle of the sine wave is completely displayed on the screen (see Fig. 51.1). Determine the number of divisions on the screen required to display exactly one cycle of the wave-

form. Multiply that number of divisions by the value for the setting of the TIME/DIV switch to obtain a value for the *period* of the sine wave (see Fig. 51.4). Compute the reciprocal of the period to obtain an experimental value for the *frequency* of the output signal from the signal generator. Your calculated frequency should be approximately 1000 Hz. Repeat this procedure for at least one other frequency setting of the signal generator.

Part 4.   Frequency Comparisons—Lissajous Figures

When two simple harmonic motions are plotted against each other at *right angles,* the resulting figure is called a Lissajous figure. Since simple harmonic motion plotted against time gives a sinusoidal waveform, if two sinusoi-

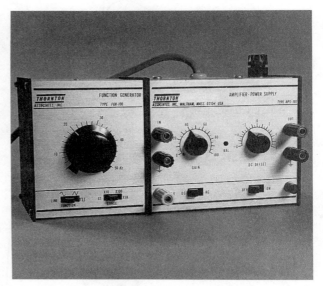

Figure 51.5. Audio oscillator (signal generator) capable of generating sine wave, triangular wave, and square wave signals with frequencies as great as 50 kHz. *(MSOE Academic Media Services)*

dal electrical signals are sent to the horizontal and vertical deflection plates of an oscilloscope, a Lissajous pattern will appear on the screen. The particular pattern depends on the frequency, amplitude, and phase relationship of the two signals.

The frequency ratio of the two signals can be determined from an analysis of the Lissajous figure produced. If the Lissajous pattern is enclosed in a rectangle as illustrated in Fig. 51.6, the frequency ratio of the two signals can be determined by counting the points of tangency along a horizontal and vertical side of the rectangle. The ratio of the frequency of the vertical input signal to the frequency of the horizontal input signal is equal to the ratio of the number of tangent points along a horizontal side of the rectangle to the number of tangent points along a vertical side (see Fig. 51.6).

1. Connect a sine-wave audio signal generator to the vertical input and the output of a 60-Hz low-voltage ac power supply to the horizontal input of the oscilloscope. If an oscilloscope with two vertical input channels is used, connect the output of the 60-Hz low-voltage power supply to the channel 1 or X input, and connect the output of the sine-wave audio generator to the channel 2 or Y input. Set the switches associated with the horizontal display so that the oscilloscope accepts the external signal from the ac power supply instead of the internally generated horizontal sweep signal. Set the signal generator for 60 Hz and make gain adjustments until a stationary ellipse of satisfactory size is observed on the screen (see Fig. 51.7). Adjust the frequency of the signal generator, and make amplitude adjustments to see if you can get a *circular* Lissajous figure.

2. Since the accuracy of the 60-Hz frequency of the ac power supply is quite dependable, that frequency will be

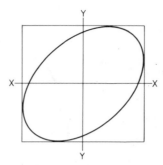

Figure 51.7. An elliptical Lissajous pattern obtained when the vertical and horizontal input signals are of the same frequency but differ in phase by about 60°. If the phase difference between the two signals were 90°, the Lissajous pattern would be a circle.

used as the reference standard for the following procedure to check the accuracy of the dial readings of the audio signal generator.

Adjust the frequency of the audio signal generator to (approximately) 30 Hz to obtain the Lissajous figure associated with a vertical-to-horizontal frequency ratio of 1:2. The Lissajous figure should look like a "figure 8." Next set the signal generator frequency to 120 Hz to obtain the 2:1 Lissajous figure that looks like a "figure 8" on its side.

In like manner, use a signal generator frequency of 40 Hz to obtain a 2:3 Lissajous figure like that in Fig. 51.6. Continue this procedure to obtain Lissajous figures for vertical-to-horizontal frequency ratios of 3:2, 5:2, 3:1, 4:1, and 5:1. Sketch all the figures obtained, and compare the frequencies determined from Lissajous ratios with the dial readings of the audio signal generator. Record these data in a neat table.

Part 5. Waveforms and Musical Sounds

Disconnect the signal generator and power supply, and return the horizontal selector switch to the internal sweep. Connect a good-quality, sensitive microphone to the vertical input.

1. Observe the waveform of the sound produced by a large tuning fork with resonating chamber. Adjust the TIME/DIV switch so that one complete wave fills the screen.

2. Strike both of the matched tuning forks (Fig. 51.8) simultaneously, and note the pattern. Adjust the TIME/DIV switch to get one full wave on the screen.

3. "Load" one of the forks to decrease its frequency by a few vibrations per second, and then strike both simultaneously. You should be able to hear the phenomenon of "beats," and also to observe it graphically on the oscilloscope screen.

4. Observe the waveforms from such typical musical instruments as the violin, guitar, clarinet, trombone, and trumpet. Make adjustments to CRO controls to "stop" the patterns.

5. Observe the waveform of the human voice at various levels of pitch (frequency). Note characteristics of consonants and vowels. If possible, note the difference in wave-

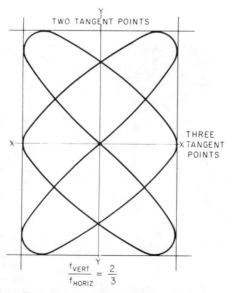

Figure 51.6. Determining the vertical-to-horizontal frequency ratio for the input signals from an analysis of the Lissajous figure formed on the oscilloscope screen. The frequency ratio is equal to the ratio of the number of tangent points along a horizontal side to the number along a vertical side of a rectangle enclosing the Lissajous pattern.

Figure 51.8. Matched tuning forks with resonating boxes. The frequency may be altered by shifting the position of the "load" on one of the bars of the fork. (*Welch Scientific Co.*)

form produced by a trained singer and that produced by your voice in singing the same note. For all of the above, provide labeled sketches of the waveforms. What is the relationship of *fundamental* tone to *overtones,* as seen on the oscilloscope screen?

DATA SHEET

Take complete notes on everything investigated as the experiment proceeds, and provide your own format for data recording. Organize your Data Sheet very carefully.

CALCULATIONS

Summarize your findings from Part 2 involving the comparison of peak-to-peak and rms voltages, your calculations from Part 3 to determine the frequency of the signal generator, and your findings from Part 4 which compare the frequencies obtained from Lissajous ratios with dial readings of the audio signal generator.

IN THE REPORT

For each step of Parts 2, 3, 4, and 5, draw sketches of the waveforms observed, and interpret the significance of the patterns. Explain how an oscilloscope can be used as an ac or dc voltmeter (Part 2) and as a frequency meter (Part 3).

Summarize what you have learned about the cathode-ray oscilloscope. From reference reading, list and describe at least five additional uses of the oscilloscope for research, production, and diagnosis in industry, business, medicine, and engineering.

NOTES, CALCULATIONS, OR SKETCHES

EXPERIMENT 52

THE BIPOLAR JUNCTION TRANSISTOR

PURPOSE

To study three different types of transistor amplifiers and a simple transistor oscillator, each constructed using a *p-n-p* bipolar junction transistor.

APPARATUS

Cenco transistor analyzer board with associated equipment (or other transistor study kit); several resistors (100Ω, $1\ k\Omega$, $10\ k\Omega$, and $100\ k\Omega$); sensitive microammeter; milliammeters with ranges of 1 and 5 mA; (digital multimeters may be used instead of the microammeter and milliammeters); two 1.5-V batteries; audio oscillator; audio interstage transformer; oscilloscope.

INTRODUCTION

The transistor is a semiconductor device first developed by the Bell Telephone Laboratories in 1948. Since 1960, the transistor has attained a position of great importance in the electronics industry, having replaced the electronic vacuum tube in nearly all applications. Although individual transistors were important circuit elements for many years, today transistors most often appear in integrated-circuit form. For example, the type 555 integrated-circuit timer used in Experiment 53 contains the equivalent of over 20 transistors. This experiment deals with the *bipolar junction transistor*.

Bipolar junction transistors are made by putting a thin section of *n*- or *p*-type silicon (or germanium) between two sections of the opposite type. (See Experiment 50 for a discussion of *n*- and *p*-type semiconductors and of *p-n* junctions.) For example, a *p-n-p* transistor is a built-up semiconductor with a thin region of *n*-type material sandwiched between two larger *p*-type regions (see Fig. 52.1). The middle section of the "sandwich" (the *base*) acts, with the application of appropriate external voltages, to control the flow of *holes* across it from the *emitter* to the *collector* regions.

Figure 52.2 shows the schematic symbol used to represent a *p-n-p* transistor in a circuit diagram; Fig. 52.3 shows the Cenco Transistor Analyzer; and Fig. 52.4 is a circuit diagram of the apparatus.

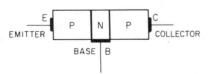

Figure 52.1. Diagram of a *p-n-p* bipolar junction transistor (BJT).

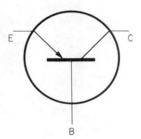

Figure 52.2. Symbol that represents a *p-n-p* bipolar junction transistor in a circuit diagram.

MEASUREMENTS

Part 1. Dc Current Gain—Common-Emitter Amplifier

A commonly used transistor amplifier circuit is one in which the input signal is applied between the emitter and the base and the output signal is taken off between the emitter and the collector. Since the emitter is common to both the input circuit and the output circuit, such an amplifier is called a *common-emitter amplifier*.

Set up the Transistor Analyzer apparatus with the common-emitter amplifier circuit as shown in Fig. 52.4. Connect 1.5-V batteries to the terminals labeled E_1 and E_2, with the positive terminal of each battery connected to the bus bar. A_1 is the microammeter, and A_2 is a milliammeter with a 5-mA range. Use a 10 kΩ resistor for R_1 and a short length of copper wire for R_2 (i.e., $R_2 = 0$). R_4 is a variable resistor.

Adjust R_4 to obtain collector currents I_C of 2, 3, 4, 5, 6, 7, and 8 mA and read the corresponding base currents I_B. Record these readings in a suitable table. Repeat the measurements using 5 kΩ for R_1.

Part 2. Ac Voltage Gain

1. With Common Emitter Connect the transistor as before (Fig. 52.4), but now connect a 6-V battery instead of the 1.5-V battery to the terminals labeled E_2. As before, connect the positive terminal of the 6-V battery to the bus bar. Use 1 kΩ for R_1 and 10 kΩ for R_2. Connect a wire shunt around A_1 and use the 1-mA range milliammeter for A_2. Adjust the variable resistor R_4 to obtain a collector current $I_C = 0.3$ mA. Connect the audio oscillator to the input terminals. Adjust it for a frequency of 1 kHz and use a low-input voltage.

Now use the oscilloscope or the digital multimeter to measure the ac output voltage for *input* voltages of 10, 20, 30, and 40 mV. Record the data for later calculations.

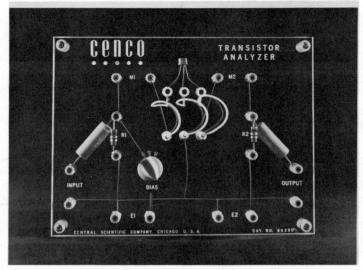

Figure 52.3. Cenco Transistor Analyzer apparatus. (*Central Scientific Co.*)

2. With Common Base Change the transistor connections on the board to accord with Fig. 52.5. (Verify for yourself that now it is the *base* which is common to both the input and the output circuits.) E_1 should be 1.5 V with *negative* terminal to the bus bar, and $E_2 = 6$ V with *positive* to the bus bar. Make $R_1 = 100\ \Omega$ and $R_2 = 100\ \text{k}\Omega$. Adjust R_4 for maximum collector current I_C.

Measure the ac output voltages for input voltages of 0.10, 0.15, and 0.20 V, with input frequency of 100 Hz. Repeat with an input frequency of 2500 Hz.

Increase the input capacitor C_1 to 1.5 μF and repeat the readings. Record all data for later interpretation.

3. With Common Collector Change the transistor connections to conform to the diagram of Fig. 52.6. (Note now that it is the *collector* which is common to both input and output circuits.) Use $E_1 = E_2 = 1.5$ V and connect *both negative terminals* to the bus bar. Make $R_1 = 10\ \text{k}\Omega$ and $R_2 = 100\ \Omega$. Meter A_1 should be shunted out, and A_2 is the milliammeter with a 5-mA range. Adjust R_4 so that the collector current I_C is about 3 mA.

Now measure the output ac voltages for input voltages

of 0.05, 0.10, 0.15, and 0.20 V at a frequency of 1 kHz. Record all data for later calculations.

Part 3. Transistor Oscillator

Make up a circuit as shown in Fig. 52.7. The *common-emitter amplifier* is merely the circuit used in Part 2, number 2, above. T is the audio interstage transformer. It may be necessary to reverse the output connections to the transformer in order to put the circuit into oscillation. An oscilloscope can be used to obtain a visual presentation of the signal at the output terminals. The oscilloscope can also be used to determine the frequency of the oscillator. Use the procedure described in Experiment 51.

CALCULATIONS

Part 1. Dc Current Gain

Plot I_C against I_B for $R_1 = 10\ \text{k}\Omega$, and compute the dc gain from the slope of the curve by evaluating the ratio $\Delta I_C/\Delta I_B$. On the same paper plot the results for $R_1 = 5\ \text{k}\Omega$. Compare the results. Interpret the meaning of the curves.

Part 2. Ac Voltage Gain

1. Common Emitter Compute the ac voltage gain (ratio of output voltage to input voltage) for each of the input voltages.

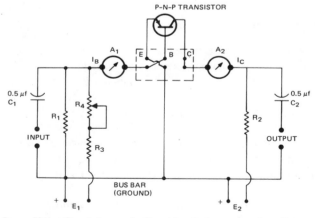

Figure 52.4. Circuit diagram for Transistor Analyzer apparatus with connections made for "common-emitter" amplifier.

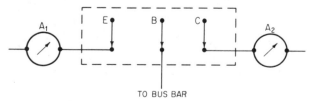

Figure 52.5. Connections for "common-base" circuit.

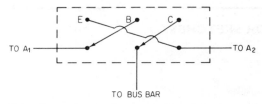

Figure 52.6. Connections for "common-collector" circuit.

2. Common Base Compute the voltage gain in the same manner for each input voltage for input signals of 100 Hz and 2500 Hz. Interpret the results.

3. Common Collector Compute the ratio of output to input potentials and state the voltage gain for each of the input potentials. Interpret the meaning of your results.

Part 3. Transistor Oscillator

Calculate the frequency of the oscillator using information from the waveform display on the oscilloscope screen.

ANALYSIS AND INTERPRETATION

1. Figure 52.2 shows the schematic symbol for a *p-n-p* transistor. Draw the schematic symbol for an *n-p-n* transistor.

2. In addition to the bipolar junction transistor (BJT), there is another type of junction transistor called a *Junction Field-Effect Transistor* (JFET). Based on some refer-

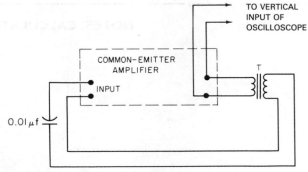

Figure 52.7. Transistor oscillator circuit.

ence reading, discuss briefly the similarities and the differences between the BJT and the JFET.

3. Are the *p-n* junctions between the emitter and base and between the base and collector forward biased or reverse biased for the common emitter amplifier? Is the answer the same for the common base and common collector amplifiers?

4. Was the frequency of the transistor oscillator in Part 3 in the range of frequencies to which the human ear is sensitive (20 Hz to 20 kHz)?

5. Based on some reference reading, discuss briefly the advantages and disadvantages of the three different types of transistor amplifiers studied in this experiment (common emitter, common base, and common collector). Mention a specific application for each type.

EXPERIMENT 53

INTEGRATED CIRCUITS—TYPE 555 IC TIMER APPLICATIONS

PURPOSE

To study three applications of the type 555 integrated-circuit timer chip and to gain experience in constructing and testing simple electronic circuits.

APPARATUS

Type 555 integrated-circuit (IC) timer chip; SPST normally open, momentary, push-button switch; resistors (1 kΩ (2), 470 kΩ, and 1 MΩ); 0–100 kΩ potentiometer; capacitors (0.1, 4.7, 22, and 47 μF); 2-1/4 inch, 8-Ω speaker; 9-V battery with snap-on connector; solderless breadboard for constructing temporary experimental circuits; approximately 60 cm (2 ft) of number 20 or 22 gauge hookup wire; soldering iron; solder; wire cutters and stripper; ohmmeter; oscilloscope, stopwatch or other timer. (All of the small electronic components listed here can be obtained at a Radio Shack store or from one of the mail-order electronic supply companies.)

INTRODUCTION

In 1958, ten years after the invention of the transistor, the integrated circuit was devised. An integrated circuit contains a number of interconnected components, such as resistors, capacitors, diodes, and transistors, formed on and within a sheet of semiconductor material approximately 0.1 mm thick with a typical area of a few square millimeters.

The first integrated circuit contained five interconnected components including one *p-n-p* transistor. By the late 1980s, integrated circuits were being manufactured that contained several million components. Compared to electronic circuits constructed with individual transistors, diodes, resistors, capacitors, and other components connected by wires or by metal paths on a printed circuit board, integrated circuits have three main advantages: They are smaller, less expensive, and more reliable. Among many other applications, the integrated circuit has made possible the modern computer industry.

The integrated circuit (IC) to be used in this experiment is the type-555 IC timer. Different manufacturers produce slightly different versions of the 555 timer, but the typical version contains 20 transistors, 15 resistors, and 2 diodes. Of the hundreds of practical applications for the type-555 IC, this experiment will involve construction of three circuits that require only a few components in addition to the integrated circuit.

MEASUREMENTS

Preparation of Components

If the components have not been previously prepared, it will be convenient to prepare them as follows before attempting to construct a circuit. Cut eight pieces of number 20 or 22 solid hookup wire to about 6-cm lengths and remove about 0.5 cm of insulation from each end of each wire. Solder one wire to the center terminal of the 100 kΩ potentiometer and one wire to either of the other terminals. Solder one wire to each of the terminals of the speaker, one wire to each terminal of the push-button switch, and one wire to each of the stranded wires connected to the snap-on connector for the 9-V battery. The solid wire will be easier to push into the holes of the solderless breadboard than the very flexible stranded wires attached to the snap-on connector.

Part I. Variable-Frequency Audio Oscillator

Figure 53.1 shows the schematic diagram for a variable-frequency audio oscillator that requires only seven components in addition to the 555 IC timer. Figure 53.2a shows a pictorial diagram of how the components can be connected on a solderless breadboard to make a working version of the circuit (see Fig. 53.2b).

After constructing the circuit, set the 100 kΩ potentiometer at about the middle of its range and push the push-button switch. If the circuit is correctly wired, you should

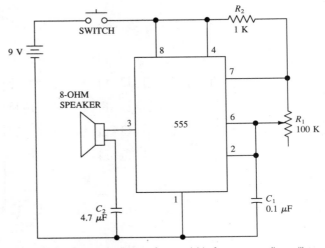

Figure 53.1. Schematic diagram for a variable frequency audio oscillator using a type-555 integrated circuit timer.

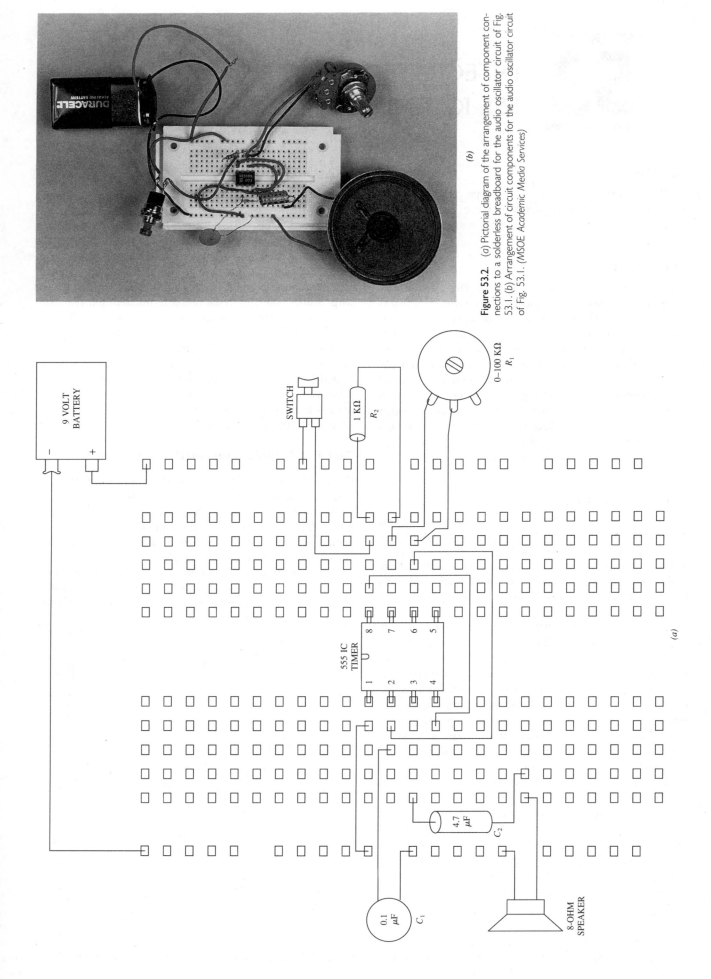

Figure 53.2. (a) Pictorial diagram of the arrangement of component connections to a solderless breadboard for the audio oscillator circuit of Fig. 53.1. (b) Arrangement of circuit components for the audio oscillator circuit of Fig. 53.1. (MSOE Academic Media Services)

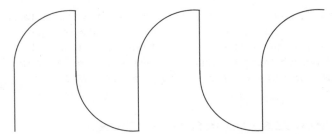

Figure 53.3. Sketch of voltage waveform across the 4.7 μF capacitor in the circuit of Fig. 53.1.

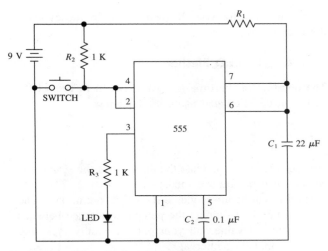

Figure 53.5. Schematic diagram of the delay timer.

hear a tone with a frequency of about 150 Hz. If you do not hear a tone, try adjusting the 100 kΩ potentiometer. If you still do not hear a tone, check your circuit for wiring errors.

Once you get the circuit operating, note the range of frequencies possible by turning the potentiometer through its range. Then set the potentiometer at about the middle of its range and connect the vertical input of an oscilloscope across the 4.7 μF capacitor (see Fig. 53.2a). The waveform displayed should look something like that in Fig. 53.3. Use the procedure described in Experiment 51 (Part 3. Measuring Frequency) to determine the period and frequency of the waveform. Use an ohmmeter to measure the resistance of the potentiometer for the last setting used, *after disconnecting the potentiometer* from the circuit. Record these results for later use.

Part 2. Variable Frequency Light-Emitting Diode (LED) Flasher Circuit

Figure 53.4 shows the schematic diagram for a circuit similar to the circuit used in Part 1, but in this case the circuit is used to cause a light-emitting diode (LED) to flash repeatedly at an adjustable rate. Use the solderless breadboard to construct the circuit as in Part 1. Note that some portions of this circuit are the same as the circuit used in Part 1, so you need not disassemble the previous circuit completely. Once the circuit is wired and the LED is flash-

ing, note that the 100 kΩ potentiometer can be used to adjust the duration of the interval between flashes. If the 100 kΩ potentiometer is set at the middle of its range, you should observe flashes of light of less than 50-ms duration with an interval of between 1 and 2 seconds between flashes.

Set the potentiometer so that the time interval between flashes is greater than 1 second, and use a stopwatch or other timer to determine an average value for the length of that time interval. Disconnect the potentiometer from the circuit and use an ohmmeter to measure the resistance between the terminals connected in the circuit. Record these values for later use.

Part 3. Delay Timer

Figure 53.5 shows the schematic diagram for a circuit that will turn on the LED for a time interval that depends on the values of R_1 and C_1. Construct the circuit first with R_1 equal to 470 kΩ. Once the circuit is wired, push the pushbutton switch and notice that the LED will turn on for between 10 and 15 s, and will then turn off. The LED will then remain off until the switch is pushed again.

Measure the average length of time the LED remains *on* after pushing the switch. Replace the 470 kΩ resistor with a 1 MΩ resistor and make the same measurement. Record these values for later use.

CALCULATIONS

Part 1. The Audio Oscillator

The frequency of the audio oscillator can be calculated using the equation (given here without derivation)

$$f = \frac{1.44}{(2R_1 + R_2)C_1} \tag{53.1}$$

Use Eq. (53.1) to calculate the frequency of the audio oscillator that you constructed in Part 1. The frequency f will be in Hz if R_1 and R_2 are in ohms and C_1 is in farads. Compute the percent difference between this calculated

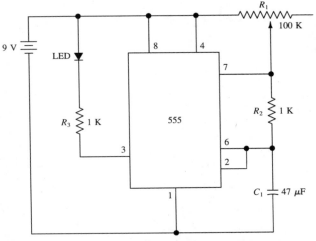

Figure 53.4. Schematic diagram of the variable-frequency light-emitting diode (LED) flasher circuit.

value and the value you obtained for the frequency using the oscilloscope.

Part 2. The LED Flasher

The duration of the time interval between flashes of the LED can be calculated using the equation

$$T = \frac{(R_1 + 2R_2)C_1}{1.44} \qquad (53.2)$$

Use Eq. (53.2) to calculate the duration of the *time interval between flashes* for the circuit of Part 2. The time interval T will be in seconds if R_1 and R_2 are in ohms and C_1 is in farads. Compute the percent difference between this calculated value and your experimentally measured value for that time interval.

Part 3. The Delay Timer

The LED in the circuit of Part 3 should remain on for a length of time T_{on} that depends on R_1 and C_1 as predicted by the equation

$$T_{on} = kR_1C_1 \qquad (53.3)$$

where k is a dimensionless constant. Use your experimen-tal data for T_{on} (in s) from Part 3, and the known values for R_1 (in ohms) and C_1 (in farads) to determine an experimental value for the constant k. Compute the percent difference between the values obtained for k from data with R_1 equal to 470 $k\Omega$ and with R_1 equal to 1 MΩ.

ANALYSIS AND INTERPRETATION

1. Based on reference reading, write a paragraph on *microelectronics* and the use of silicon wafer-chips in the manufacture of integrated circuits.

2. Suggest practical applications for each of the three circuits constructed in this experiment. Consider the possibility of revising the circuits slightly or of using them in conjunction with other circuits.

3. For the audio oscillator circuit of Part 1, what values of components could be used for an oscillator frequency of 100 Hz?

4. For the LED flasher circuit of Part 2, what values of components could be used so that the *time interval between flashes* would be 1.0 s?

5. For the delay timer circuit of Part 3, use your experimental value for k to determine the value of R_1 needed to make T_{on} equal to 10 s.

EXPERIMENT 54

RADIOACTIVE DECAY—HALF-LIFE DETERMINATION

PURPOSE

To study the decay rate of a radioactive isotope, and to determine its half-life.

APPARATUS

Cesium/barium minigenerator kit (Union Carbide or equivalent); Geiger tube, amplifier-power supply, decade counter, and preset timer (all Thornton or equivalent); planchet; laboratory time clock; stands and clamps.

INTRODUCTION

A substance is designated as radioactive if its nucleus is unstable and spontaneously emits *alpha* (α), *beta* (β), or *gamma* (γ) radiation. Such substances occur in nature, usually as compounds of elements near the "heavy end" of the periodic table. Artificially radioactive substances can be prepared also, using the methods of "high-energy" physics.

Natural atomic disintegration was first observed by Becquerel and by Pierre and Marie Curie, near the close of the nineteenth century. They had been studying certain ores of uranium for some years, and had noted that three types of radiations were emitted spontaneously from these materials. Shortly afterwards, the element *radium* itself was isolated and identified by Marie Curie, and intensive studies were made of its emanations.

These radiations were carefully studied by Rutherford in 1903, and their energies were determined. Three separate forms were identified—alpha (α), beta (β), and gamma (γ), named for the first three letters of the Greek alphabet. It was subsequently found by other investigators that the gamma rays were *photons* of very short wavelength (electromagnetic radiation). Alpha "rays" were shown to be helium nuclei, and beta "rays" proved to be high-speed electrons ejected from atomic nuclei. These discoveries were the first indication of the problem of trying to distinguish between particles and waves—the so-called *wave-particle duality* of modern physics.

Alpha, beta, and gamma rays differ in their power or ability to penetrate matter and also in their ability to produce ionization. Their behavior in the presence of a magnetic field is also widely different. A simple experiment has been devised which makes use of this behavior in a magnetic field.

As shown in Fig. 54.1, a small amount of radioactive material is placed at the bottom of a hole drilled in a lead block. The block and the photographic plate are placed a few centimeters apart in an airtight chamber, which is then

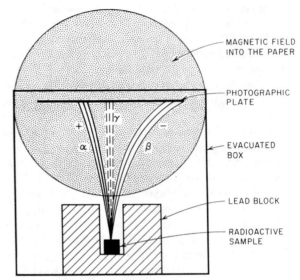

Figure 54.1. Diagram illustrating the differing behavior of alpha, beta, and gamma radiations in the presence of a magnetic field.

evacuated. A strong magnetic field is then applied at right angles to the emerging stream of radiations. After a reasonable "exposure" time, the photographic plate is developed, and three separate exposed spots are found. The alpha particles, being deflected to the left, produce one spot; the beta particles are deflected even more markedly to the right; and the gamma rays are not deflected at all. From the known direction of the magnetic field, it may be deduced that the alpha particles are positively charged and that the beta particles are negatively charged. Since the gamma rays are not deflected by the magnetic field, they must carry no electrical charge.

Alpha particles, although they are good ionizers (since they carry a double positive charge), do not have high penetrating power. They are absorbed by a few centimeters of air or by aluminum foil about 0.006 cm thick or by a sheet of typing paper. *Beta particles* have less ionizing power than alpha particles, but are about 100 times more penetrating. A sheet of aluminum almost 3 mm thick is required to absorb them. *Gamma rays,* having no electrical charge, produce very little ionization, but they have tremendous penetrating power, being able to pass through a block of iron about a foot thick.

It has been determined that *alpha particles* are actually helium nuclei. Their mass is about four times that of a hydrogen atom, and each particle carries a double positive charge. Alpha particles are emitted at velocities on the order of one-twentieth the speed of light.

Beta particles have the same mass and charge as electrons and *are* regarded as being electrons. Their velocity is extremely high, approaching that of light.

Gamma rays are not charged particles, but rather are photons of electromagnetic radiation of extremely short wavelength, shorter than most x-rays.

Geiger Tube Radiation Detector   The Geiger tube consists of a metal tube, usually sheathed inside a glass cylinder with a very fine tungsten wire suspended axially along the tube (see Fig. 54.2). Argon gas at a pressure of about 10 cm Hg fills the tube. A potential difference of 900 to 1000 V is applied between the tungsten wire and the metal cylinder, positive (+) to the wire, and negative (−) to the cylinder. This high potential difference is just slightly less than that required to produce an electrical discharge within the tube.

When an alpha or beta particle or a gamma ray photon enters the tube, it produces ions from the gas atoms in the tube. The electric field in the tube accelerates these ions, and more ions are produced by collision. Thus, an ionization current is rapidly built up (in less than a millionth of a second). This ionization current is sufficient to "trigger" the tube, and a transient current flows in the external circuit, dissipating itself to ground. The tube is almost immediately ready for another entering particle. A Geiger tube is most effective at detecting beta particles. It is less effective at detecting alpha particles because they are so easily absorbed by the thin "end window" where radiation enters the tube. The Geiger tube is less effective at detecting gamma rays because their penetrating power allows most of them to pass through the gas in the tube without interacting with the gas atoms to produce the ions needed to trigger the tube. The tube may be connected externally through an amplifier to a neon indicator lamp and a loudspeaker. It may also be connected to a count-rate meter, or a decade counter. The amplifier and speaker make audible the passage of a particle or ray through the tube.

Radioactive Isotopes   *Isotopes* are different forms of the same basic element. The element barium is made up of different kinds of atoms, some of them being ^{132}Ba, ^{134}Ba, and ^{137}Ba. The number preceding the chemical symbol *Ba* gives the total number of *protons and neutrons* in the nucleus. Barium has an atomic number (Z) of 56,

which is the number of protons in the nucleus of all barium atoms, including all the isotopes of barium. Therefore, by subtracting, we find that ^{132}Ba has 76 neutrons, ^{134}Ba has 78 neutrons, and ^{137}Ba has 81 neutrons in the nucleus. ^{137}Ba is a *radioactive isotope,* which means that it is unstable, and gives off radiations as it changes form or *decays.*

Each radioactive isotope has its own unique rate of decay, called its *half-life.* Half-life is the time it takes for 50 percent of a given amount of the isotope to change form, or the time required for the radiation rate to drop to one-half of whatever it was at the beginning of the time interval. Half-life varies widely from isotope to isotope, examples being Uranium 238 with a half-life of 4.5×10^9 years; Radium 226 with a half-life of 1622 years; and Polonium 218 with a half-life of three minutes. The isotope you will use (^{137}Ba) has a half-life of less than five minutes.

Barium 137 emits a high-energy, very-short-wavelength gamma ray each time an atom decays. By use of the Geiger tube, the emitted ray can be detected, thereby verifying the fact that an atom has changed form. (The Geiger tube can detect only a certain percentage of the total number of emitted gamma rays from the sample. However, if the *detection rate* decreases with time the decrease is due to the isotope's decreasing its activity, not to a changing percentage detected by the Geiger tube.)

The decade counter electronically records the number of signals from the Geiger tube. By setting the pre-set timer to a desired value, you can keep track of the activity of the isotope, measured in "counts" per time interval. As time passes, the count rate will drop as the ^{137}Ba decays. By plotting counts versus time, the half-life of the ^{137}Ba isotope can be obtained (see Fig. 54.3).

MEASUREMENTS

Mount the Geiger tube about 10 cm above the (empty) planchet, and proceed to measure the *background radiation*. This is a radiation level due to random radiations present at your location from natural causes. To do this set the pre-set timer at 250 seconds and push the "Start" button. After 250 seconds, the counting system will shut off. Record the counts. Divide by 25 to obtain the background count in *counts per 10 seconds*. (See Fig. 54.4 for a typical arrangement of equipment.)

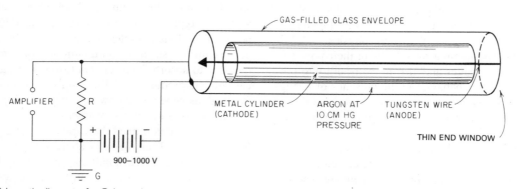

Figure 54.2.   Schematic diagram of a Geiger tube.

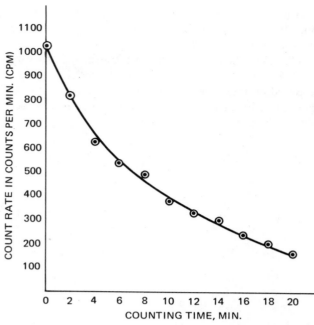

Figure 54.3. Half-life curve for a hypothetical radioactive isotope (not ^{137}Ba), with counts taken every two minutes. Note that the count rate is reduced to one-half the initial value after approximately 8 minutes, and to one-fourth the initial value after approximately 16 minutes. The half-life for this isotope is therefore approximately 8 minutes. Your data probably will not be as "perfect" as this illustrative plot suggests. Note that the data plotted here are in counts per minute, while your data will be in counts per 10 seconds.

Now take the *planchet* to the instructor. You will be shown how to use the minigenerator kit to obtain the ^{137}Ba isotope[1] (see Fig. 54.5). Place the isotope (in solution) in the planchet and bring it back at once to your counting station, place it under the Geiger tube, and begin taking counts immediately, as follows.

Change the preset timer to 10 seconds. The counting process will be to count for 10 seconds, rest for 10 seconds, count 10, rest 10, etc. Push the "Start" button to begin the first 10-second count. During the 10-second rest period, record the counts and *reset the decade counter*. On this schedule, you will be starting counts when the laboratory clock reads 0, 20 seconds, 40 seconds, 60 seconds, etc. Be sure you understand this procedure, because a slip-up during the process will necessitate starting over from the beginning. The half-life is quite short for your ^{137}Ba sample, and any appreciable delay means you have lost a high percentage of your sample. Proceed with the above counting exercise *immediately* after obtaining your isotope sample.

Continue counting until the recorded count for 10 seconds is *less than one-fourth the initial count* (in the first 10-second period). The isotope is now only about one-fourth its original strength. Place the planchet in a tray provided by the instructor. Do not pour out the solution!

<hr>

[1]The radioactive test sample (^{137}Ba) is not dangerous, but do not handle it unnecessarily, or spill any on your skin or clothing. Always observe the utmost precautions in the presence of radioactive materials!

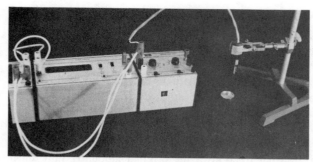

Figure 54.4. Equipment setup for counting. The decade counter is at left, then the amplifier-power supply for the Geiger tube. The Geiger tube is shown suspended over the planchet containing the radioactive sample, at right. (*Thornton Associates*)

Any spillage should be wiped up immediately. At the conclusion of the experiment, wash your hands thoroughly.

CALCULATIONS AND RESULTS

Subtract the background count from each recorded ^{137}Ba count. This gives the corrected count in counts per 10 seconds, due to the radioactive sample only.

Plot corrected counts, versus time on graph paper. Notice that the time scale will be in multiples of 20 seconds. Plot the points so that the corrected counts are at 5 seconds, 25 seconds, 45 seconds, etc. Your data represent an *average* count rate for each 10-second interval, and so the count rates you found are more typical of the midpoints of the time intervals rather than the starting or end points.

Draw a "smoothed" curve between the plotted points. Determine the time at which the count rate was exactly one-half the initial count rate. Then the time at which the count rate was exactly one-fourth the initial value. By subtraction, find first, the time taken for the count to fall off to one-half its original value, and second, the time required for the 50 percent count to fall off to one-half its value, or 25 percent of the original. (See Fig. 54.3.)

Average the two times found to get your best estimate for the half-life of ^{137}Ba. Compare your result with the accepted value for ^{137}Ba half-life, available from your instructor. Compute your percent error.

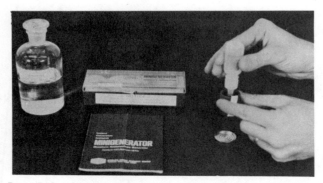

Figure 54.5. Cesium-barium minigenerator kit, for obtaining the radioactive ^{137}Ba isotope. (*Union Carbide*)

ANALYSIS AND INTERPRETATION

1. Are your two estimates of half-life essentially the same? Should they be? Would a larger count rate improve the agreement?

2. Was background radiation a crucial factor in this experiment?

3. If the counting were increased to 20 seconds, how would the results have been affected, if at all?

4. A radioactive isotope has a half-life of 20 h. After 80 h, what fraction of the original amount is left undecayed?

5. A reactor produces 256 μg of a certain radioactive isotope. If only 16 μg remains six days later, what is the half-life of the isotope?

6. Why must the end window for a Geiger tube be very thin if it is to successfully detect alpha particles?

7. Based on reference reading, write a paragraph to explain how *radioactive dating methods* can be used to determine the age of very old objects.

EXPERIMENT 55

THE WILSON CLOUD CHAMBER— PARTICLE TRACKS

PURPOSE

To study the operation of a diffusion (Wilson) cloud chamber, and to observe typical particle and ray tracks therein.

APPARATUS

Diffusion cloud chamber (Fig. 55.1), with clearing field; sources of alpha, beta, and gamma radiation; slab of dry ice; methyl alcohol; laboratory light source of high intensity; 45-V dc battery to supply the clearing field for the cloud chamber.

INTRODUCTION

Natural radioactivity goes on around us all the time. Radioactive ores in their normal decay processes give off *alpha* and *beta* particles and *gamma rays*. High-energy particles from outer space known as *cosmic rays* are "pelting" us continually. Atomic and nuclear reactions produce high-speed *electrons* and *neutrons*, and so-called "secondary reactions" produce a variety of atomic spare parts and emanations such as *protons, mesons, gamma rays*, and *x-rays*.

These particles and rays cannot be felt, seen, heard, or sensed by the human body in the normal course of events. Certain scientific devices have been developed, however,

which enable us to detect their passage. The Geiger tube (studied in Experiment 54) is one of these. Another is the *Wilson cloud chamber*.

The cloud chamber was devised by C. T. R. Wilson in England in 1912. He was awarded the Nobel Prize for physics in 1927 on the basis of the research he carried on with this instrument.

The cloud chamber has been one of the atomic scientist's most useful tools, and many of the elementary particles of matter were first identified by its use. The first observation of a nuclear transmutation was made with a cloud chamber.

The condition of the vapor in a cloud chamber must be "just right" before particle "tracks" can be observed. The Wilson and other early cloud chambers produced the proper saturated vapor conditions for track formation by a sudden expansion of air trapped in a chamber over a water reservoir (adiabatic cooling). A transient condition (lasting about 1/25 second) favorable to track formation resulted.

The cloud chamber used in this experiment, however, makes use of dry ice as a coolant; methyl alcohol vapor, and a charging electric field. A supersaturated atmosphere which is favorable to ray track formation can be maintained for more than an hour.

Figure 55.1. Diffusion cloud chamber with high-intensity light source, block of dry ice, battery to establish the clearing field, and radioactive samples. *(Atomic Laboratories, Inc., Berkeley)* (The Central Scientific Co. also supplies a similar apparatus that gives excellent results.)

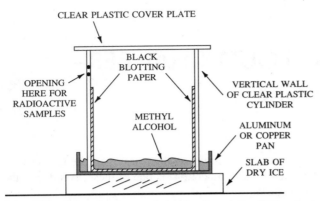

Figure 55.2. Sketch showing a vertical section through the diffusion cloud chamber. Blotting paper should cover the bottom and about two-thirds of the way up the sides of the lucite cylinder.

CLEAR PLASTIC COVER PLATE

BLACK BLOTTING PAPER

OPENING HERE FOR RADIOACTIVE SAMPLES

VERTICAL WALL OF CLEAR PLASTIC CYLINDER

METHYL ALCOHOL

ALUMINUM OR COPPER PAN

SLAB OF DRY ICE

What is observed in a cloud chamber is *not* the particle or the ray itself, but the path that it takes through the chamber. In an atmosphere which contains more vapor than it normally should for the existing temperature and pressure conditions (the condition of *supersaturation*), any foreign object may serve as a starting point for the growth of a visible droplet of fog or mist. Dust particles often serve this purpose in the earth's atmosphere, and fog or clouds form. The passage of a jet plane at high altitudes causes *contrails* (*con*densation *trails*) of condensed vapor to form. Charged particles and/or ions have the same effect. Neutrons and gamma rays carry no charge and cause no ionization. They do, however, through collision, produce charged *secondaries* whose "tracks," made up of fog droplets, may be observed in the cloud chamber. These particle (or ray) tracks are transitory—visible only momentarily—and can only be seen under optimum lighting conditions.

With the cloud chamber then, high-energy particles and rays, though they themselves cannot be seen by any known technique or instrument, can be studied in terms of the fog or "cloud" tracks they form as they collide with atoms and form ions.

The student cloud chamber to which the following directions apply (Figs. 55.1 and 55.2) consists of a clear plastic (lucite) cylinder sitting in a shallow metal pan. The top of the lucite cylinder is covered by a clear plastic plate. The metal pan at the bottom, and the interior surface (2/3 of the way up) of the clear plastic cylinder are to be covered with black blotting paper, which serves two purposes: (1) It provides an excellent background against which the illuminated fog tracks can be observed, and (2) it absorbs methyl alcohol from the reservoir in the metal pan and provides a large surface area from which the alcohol can evaporate, cooling and filling the cylinder with saturated vapor. The entire chamber and metal pan sit on a slab of dry ice. There is a port, or window, in the cylinder wall for insertion of the radioactive sources. A strong light is positioned so that the inside of the chamber is brightly illuminated. Provision is made for the establishment of a clearing electrostatic field.

METHOD AND OBSERVATIONS

Set up the cloud chamber in accordance with the instructor's and the manufacturer's directions.[1]

Carefully wash and dry the cover glass. It must be very clean.

Fill the bottom of the chamber with methyl alcohol to a depth of about ¼ in. Cover the bottom and part way up the sides with black blotting paper. A perfectly black background makes the ray tracks more easily visible. Be sure the alcohol level is such that it touches the bottom of the black blotter which serves as the inner liner of the chamber.

Replace the cover plate, making sure that the aluminum ring (not shown in Fig. 55.2) fits snugly over the top.

Place the chamber on a slab of dry ice about 2 cm thick and 15 cm square.

Attach the positive terminal of the clearing field dc source (45 to 90 V) to the bottom tray, and the negative terminal to the top aluminum ring which makes contact with the electrode on the underside of the plastic cover.

Place the high-intensity light source (ordinary incandescent lamps or a flashlight are not satisfactory) about 15 cm from the open space (port) in the chamber side wall, with its beam directed so that it glances off the alcohol surface in the bottom of the chamber.

Allow to stand about 10 to 20 min for the cloud chamber to cool sufficiently to establish the condition of supersaturation for methyl alcohol vapor. The spotlight should be turned off during this time to avoid heating of the interior of the cloud chamber. The "clearing field" voltage is applied in order to sweep out ionized particles caused by normal "background" radiations, such as cosmic rays.

Part I. Cosmic-Ray Tracks

After operating conditions have been achieved, observe the cloud chamber without any source of radiation nearby. In this situation any tracks seen will probably be due to *cosmic rays*. These ray tracks are thin and somewhat faint. They may be straight but are more usually crooked or "wiggly." The sketches of Fig. 55.3 show typical cosmic-ray tracks. Observe these tracks for several minutes and make sketches of the several patterns you see.

Rarely, you may see a "cosmic-ray shower." This event occurs when a high-energy cosmic ray interacts with

---

[1]The directions provided here apply to the *Atomic Laboratories* cloud chamber illustrated in Fig. 55.1. If a different cloud chamber is used, follow the manufacturer's instructions.

Figure 55.3. Typical cosmic ray tracks from natural background radiation.

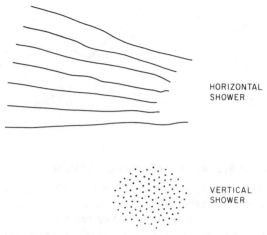

HORIZONTAL
SHOWER

VERTICAL
SHOWER

Figure 55.4. Cosmic ray "showers."

Figure 55.6. Alpha-particle tracks produced by particles from an alpha source.

atoms in the air within the cloud chamber or with atoms of the walls of the chamber. The probable explanation is that a high-energy electron in passing through the coulomb field of a heavy nucleus causes the emanation of a "shower" of gamma rays. Each gamma ray formed "dies" while liberating an electron and a positron in a process called *pair production*. Thus a cascade or "shower" of tracks is produced in the cloud chamber. Look for the patterns as shown in Fig. 55.4.

Part 2. Alpha-Particle Tracks

Alpha particles are continually emanating from the gas *radon,* which is present in very small quantities in the air and in the earth's crust. Alpha particles are actually helium nuclei, and they carry a double positive charge. They are therefore heavily ionizing and make broad, easily recognized tracks in the cloud chamber. The tracks have a range which averages about 4 cm. They sometimes have a "hook" at the end. The disintegration of the track into a "puff" at the end is a phenomenon also frequently observed.

First, see if you can find some alpha-particle tracks from natural radon disintegration. Look for the patterns shown in Fig. 55.5.

Next place the alpha source within the cloud chamber. (*Note:* Handle all radioactive materials with extreme care!

Keep them away from the face and body. Wash your hands thoroughly when the lab period is over.) Now all the tracks will start from the source, and there should be many of them to observe. Look for a pattern like that in Fig. 55.6. Alpha tracks from an alpha-particle source will undergo some self-absorption, and the tracks will not have a definite range, some being much shorter than others.

Observe alpha tracks for several minutes, and make careful sketches of what you see.

Part 3. Beta-Particle Tracks

The tracks you observed in Step 1 as a result of cosmic radiation were mostly beta tracks. Beta "rays" are actually high-energy electrons. Electrons, having unit charge, are not as good ionizers as alpha particles (helium nuclei), with their double charge. Beta tracks are therefore generally thinner and "weaker" than alpha tracks. Placing a beta source in the chamber produces many beta tracks for observation. They will be somewhat straighter and better defined than those produced by betas from normally existing cosmic radiation.

Look for a pattern like that in Fig. 55.7.

Sketch carefully the result of your observations.

Part 4. Gamma Rays and Neutrons

Gamma rays and neutrons leave no tracks (of their own) in cloud chambers. Since they are electrically neutral, they cause no ionization and therefore no train of condensed fog particles. They may (and frequently do) produce charged "secondaries" (photoelectrons and recoil protons) whose tracks are observable.

By placing a gamma source outside and near the chamber, one may observe what appears to be cosmic showers from the direction of the source. See if you can set up this condition. Sketch what you observe.

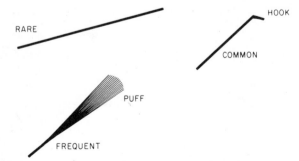

RARE

HOOK

COMMON

PUFF

FREQUENT

Figure 55.5. Sample alpha-particle tracks from natural (radon) disintegration.

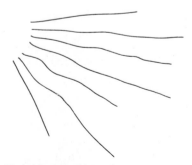

Figure 55.7. Typical beta-particle tracks.

Part 5. Nuclear Interactions

Rarely, you may see some cloud-chamber tracks which result from nuclear interactions. Nuclear interactions occur as the result of such events as particles (alphas or protons) striking gas atoms and knocking one or more particles out of the nucleus of the atom.

Look for patterns like those in Fig. 55.8.

Sketch any such patterns that you see.

DATA SHEET

There is no Data Sheet. The observations, notes, and sketches constitute the data for this experiment.

PRESENTATION OF RESULTS

There are no calculations for the experiment. All sketches should be re-done carefully for the final report, and each should be accompanied by a full explanation of what was done, and what was observed. Label each track or set of tracks with an identification of its source, according to your best judgment.

Figure 55.8. Sample of tracks resulting from nuclear interactions.

ANALYSIS AND INTERPRETATION

1. High-energy (particle) physics is a rapidly expanding field of investigation today. From your textbook and other sources, obtain information and submit a brief discussion of such particles as: *neutrinos, muons, strange particles, quarks,* and *mesons*.

2. What are some of the characteristics and probable capabilities of the *Super-Collider,* which is currently (1989) in the preliminary design stage? (Ask your instructor and the college librarian for sources of information.)

APPENDIX

TABLE I.   ENGLISH AND METRIC UNITS AND EQUIVALENTS _____

1 inch (in.) = 2.54 centimeters (cm)
1 foot (ft) = 30.48 cm = 0.3048 meter (m)
1 mile (mi) = 1.61 kilometers (km) = 5280 ft
1 microinch (μin.) = 10^{-6} in.

1 cm = 0.3937 in.
1 m = 39.37 in. = 3.28 ft
1 km = 1000 m = 0.621 mi
1 m = 100 cm = 1000 millimeters (mm)
1 micron (μ) = 10^{-6} m (also called micro-meter)
1 nanometer (nm) = 10^{-9} m
1 angstrom (Å) = 10^{-10} m

1 pound (lb) = 453.6 grams-force = 4.45 newtons (N)
1 lb = 0.454 kilograms-force (kg-f)
1 lb = 7000 grains (gr)
1 slug (sl) = 32.17 lb = 14.59 kg
1 lb (force) = 4.45 N
1 ton (force) = 2000 lb

1 kilogram (force) = 9.81 N = 2.205 lb = 1 "kilo"
1 N = 0.102 kg-f = 0.225 lb (force)

1 gallon (gal) = 231 cubic inches (in.^{3})
1 gal = volume of 8.34 lb of water

1 liter (L) = 1.057 quarts (qt) = 61.02 in.^{3}

TABLE II.   PHYSICAL CONSTANTS _____

Acceleration of gravity g	32.17 feet per second per second (ft/s^2) 981 cm/s^2, or 9.81 m/s^2
1 standard atmosphere	14.7 lb/in.^{2}, or 1.013×10^5 pascals (Pa) 29.95 in. of mercury, or 760 torrs 76 cm of mercury, or 33.94 ft of water 1 pascal (Pa) = 1 newton/meter^2(N/m^2)
Absolute zero	−273.15 degrees Celsius (°C) −459.7 degrees Fahrenheit (°F) 0 degree Rankine (°R) 0 Kelvin (K)
Mechanical equivalent of heat J (Joule's constant)	4.186 joules per calorie (J/cal), or 4186 J/kcal 778 foot-pounds per Btu (ft•lb/Btu)
Velocity of sound in air at: 0°C 32°F	331 meters per second (m/s) 1087 feet per second (ft/s)
Variation of sound velocity with air temperature	+1.1 ft/s per F° temperature rise +0.6 m/s per C° temperature rise
Velocity of light in vacuum	2.998×10^8 m/s 186,000 mi/s
Ice point of water, 1 atm	0°C = 273K 32°F = 492°R
Boiling point of water, 1 atm	100°C = 373K 212°F = 672°R

TABLE III. USEFUL NUMERICAL RELATIONSHIPS

Geometrical Figures

Circle:

Circumference	$C = 2\pi r$
Area	$A = \pi r^2$

Sphere:

Volume	$V = 4/3\pi r^3$
Surface area	$A = 4\pi r^2$

Right Circular Cylinder:

Volume	$V = \pi r^2 h$
Total surface area	$A = 2\pi r(h + r)$

Right Circular Cone:

Volume	$V = 1/3\pi r^2 h$

Angle Measurement

1 revolution = 360 degrees
1 radian = 57.3 degrees
1 revolution = 2π radians
90 degrees = $\pi/2$ radians
1 degree = 0.01745 radian = 60 min (arc)
1 degree = 3600 sec (arc)

TABLE IV. DENSITIES

Solids

Material	Mass density (g/cm^3)	Weight density (lb/ft^3)	Material	Mass density (g/cm^3)	Weight density (lb/ft^3)
Aluminum	2.70	168	Silver	10.5	655
Brass (70% Cu, 30% Zn)	8.45	540	Steel	7.8	486
Copper (annealed)	8.88	555	Tungsten	18.8	1170
Glass (common)	2.4 to 2.6	150 to 160			
Gold	19.3	1204	Woods:		
Ice	0.91	57	Cork	0.2 to 0.3	...
Iron	7.87	490	Maple	0.6 to 0.9	...
Lead	11.36	710	Oak	0.6 to 0.9	...
Platinum	21.5	1340	Pine	0.4 to 0.7	...

Liquids, 4°C

Material	Mass density (g/cm^3)	Weight density (lb/ft^3)	Material	Mass density (g/cm^3)	Weight density (lb/ft^3)
Alcohol (ethyl)	0.794	49.4	Turpentine	0.873	54.3
Benzene	0.90	56.1	Water (pure, 4°C)	1.00	62.4
Carbon bisulfide	1.29	80.7	Water (ocean)	1.025	64
Carbon tetrachloride	1.60	99.6	Gasoline	0.68	42.0
Kerosene	0.82	51.2	Mercury	13.6	849

Gases
(1 atm pressure, 0°C; 32°F)

Material	Mass density (g/cm^3)	Weight density (lb/ft^3)	Material	Mass density (g/cm^3)	Weight density (lb/ft^3)
Air (dry)	0.001293	0.0807	Carbon dioxide	0.00197	0.1234
Helium	0.000178	0.0111	Hydrogen	0.0000897	0.0056
Methane	0.0103	0.6448	Nitrogen	0.001255	0.781
Oxygen	0.00143	0.0892			

TABLE V. YOUNG'S MODULUS Y

Material	N/m^2	$lb/in.^2$
Aluminum	7.0×10^{10}	10.2×10^6
Brass, drawn	13.0×10^{10}	19×10^6
Copper	12.5×10^{10}	18×10^6
Iron (wrought)	18.5×10^{10}	26×10^6
Steel (drawn)	20×10^{10}	29×10^6

TABLE VI.   SHEAR MODULUS N

Material	N/m^2	$lb/in.^2$
Aluminum	2.5×10^{10}	3.64×10^6
Brass	3.7×10^{10}	5.37×10^6
Copper	4.5×10^{10}	6.1×10^6
Steel (mild)	8.0×10^{10}	11.6×10^6

TABLE VII.   WORK, POWER, AND HEAT UNITS

$$1 \text{ foot-pound (ft·lb)} = 1.356 \text{ joules (J)}$$
$$1 \text{ horsepower (hp)} = 33,000 \text{ ft·lb/min}$$
$$= 550 \text{ ft·lb/s}$$
$$= 746 \text{ watts (W)} = 0.746 \text{ kilowatt (kW)}$$
$$1 \text{ watt} = 1 \text{ J/s} = 3.4 \text{ Btu/h}$$
$$1 \text{ kilowatt} = 1000 \text{ watts} = 1.34 \text{ hp}$$

$$1 \text{ Btu} = 252 \text{ calories} = 0.252 \text{ kcal} = 1055 \text{ J} = 778 \text{ ft·lb}$$
$$1 \text{ calorie (cal)} = 4.186 \text{ J}$$
$$1 \text{ kilocalorie (kcal)} = 4186 \text{ J}$$
$$1 \text{ joule (J)} = 0.239 \text{ cal} = 9.48 \times 10^{-4} \text{ Btu}$$

$$\text{Latent heat of fusion of water } L_f = 144 \text{ Btu/lb}$$
$$(\text{at } 0°C \text{ or } 32°F) = 80 \text{ cal/g}$$
$$= 80 \text{ kcal/kg}$$

$$\text{Latent heat of vaporization of water } L_v = 970 \text{ Btu/lb}$$
$$(\text{at } 1 \text{ atm pressure, } 212°F \text{ or } 100°C) = 540 \text{ cal/g}$$
$$= 540 \text{ kcal/kg}$$

TABLE VIII.   COEFFICIENTS OF LINEAR EXPANSION

Material	Per Celsius Degree	Per Fahrenheit Degree
Aluminum	24×10^{-6}	13×10^{-6}
Brass	18×10^{-6}	10×10^{-6}
Copper	17×10^{-6}	9.5×10^{-6}
Glass	9×10^{-6}	5×10^{-6}
Iron (steel)	12×10^{-6}	6.7×10^{-6}
Lead	29×10^{-6}	16×10^{-6}
Silver	19×10^{-6}	10.5×10^{-6}

TABLE IX.   SPECIFIC HEAT (cal/g·C° OR kcal/kg·C° OR Btu/lb·F°)

Air	0.24
Aluminum	0.22
Brass	0.092
Copper	0.093
Glass	0.20
Ice	0.51
Iron (steel)	0.115
Lead	0.03
Mercury	0.033
Steam	0.48
Water	1.00

TABLE X.   RESISTIVITY OF METALS

	Resistivity (18 °C)	
Elements	(in ohm-circular mil/ft)	(in ohm-m)
Aluminum	16.9	2.8×10^{-8}
Copper	10.3	1.7×10^{-8}
Gold	14.0	2.4×10^{-8}
Iron	72 to 84	12 to 14×10^{-8}
Lead	132	22×10^{-8}
Silver	9.7	1.6×10^{-8}
Tungsten	33	5.5×10^{-8}

	Resistivity (18 °C)	
Alloys	(in ohm-circular mil/ft)	(in ohm-m)
Brass	38 to 52	6.3 to 8.6×10^{-8}
Constantan (Cu 60%, Ni 40%)	295	49×10^{-8}
German silver	199	33×10^{-8}
Nichrome (Ni 60%, Cr 12%, Mn 2%, Fe 26%)	660	110×10^{-8}

TABLE XI.   ELECTRICAL AND ELECTROMAGNETIC RELATIONSHIPS

$$1 \text{ ampere (A)} = 1 \text{ coul per second (C/s)}$$
$$1 \text{ coulomb (C)} = 1 \text{ A} \cdot \text{s}$$
$$= 1 \text{ farad} \cdot \text{volt}$$
$$1 \text{ ohm } (\Omega) = 1 \text{ volt/amp (V/A)}$$
$$1 \text{ farad (F)} = 1 \text{ coul/volt (C/V)}$$
$$1 \text{ microfarad } (\mu\text{F}) = 10^{-6} \text{ farad (F)}$$
$$1 \text{ picofarad (pF)} = 10^{-12} \text{ F}$$
$$1 \text{ volt (V)} = 1 \text{ joule/coulomb (J/C)}$$
$$1 \text{ henry (H)} = 1 \text{ volt} \cdot \text{sec/amp (V} \cdot \text{s/A)}$$
$$= 1 \text{ ohm} \cdot \text{sec}$$
$$= 10^{3} \text{ millihenrys (mH)}$$
$$1 \text{ joule (J)} = 1 \text{ volt} \cdot \text{coul (V} \cdot \text{C)}$$
$$= 1 \text{ weber} \cdot \text{amp (Wb} \cdot \text{A)}$$
$$1 \text{ electron volt (eV)} = 1.602 \times 10^{-19} \text{ J}$$
$$1 \text{ MeV} = 1.602 \times 10^{-13} \text{ J}$$
$$1 \text{ coulomb (C)} = 6.24 \times 10^{18} \text{ electron charges (e)}$$
$$1 \text{ ampere} \cdot \text{hour (A} \cdot \text{hr)} = 3,600 \text{ C}$$
$$1 \text{ weber (Wb)} = 1 \text{ N} \cdot \text{m/A} = 1 \text{ J/A}$$
$$1 \text{ Wb/m}^2 = 1 \text{ tesla (T)} = 1 \text{ kg/C} \cdot \text{s}$$
$$1 \text{ Wb} = 1 \text{ T} \cdot \text{m}^2$$
$$1 \text{ N/Wb} = 1 \text{ A/m}$$
$$1 \text{ T} = 1 \text{ N/A} \cdot \text{m}$$